DC

Design Cognition
——Cognitive Science in Design

设计认知
——设计中的认知科学

陈超萃
艾奥瓦州立大学
建筑系及人机接口研究组教授

Chiu-Shui Chan
Professor of Architecture/HCI
Department of Architecture/Virtual Reality Applications Center
Iowa State University

中国建筑工业出版社

图书在版编目（CIP）数据

设计认知——设计中的认知科学/陈超萃编著.—北京：中国建筑工业出版社，2008
ISBN 978-7-112-09897-2

Ⅰ.设… Ⅱ.陈… Ⅲ.建筑设计—关系—认知科学—研究 Ⅳ.TU2-05

中国版本图书馆 CIP 数据核字（2008）第 017753 号

责任编辑：陈　桦　吕小勇
责任设计：董建平
责任校对：兰曼利　刘　钰

设计认知——设计中的认知科学
Design Cognition——Cognitive Science in Design
陈超萃
艾奥瓦州立大学
建筑系及人机接口研究组教授
Chiu – Shui Chan
Professor of Architecture / HCI
Department of Architecture / Virtual Reality Applications Center
Iowa State University

*

中国建筑工业出版社出版、发行（北京西郊百万庄）
各地新华书店、建筑书店经销
北京永峥印刷有限责任公司制版
北京富生印刷厂印刷

*

开本：889×1194 毫米　1/20　印张：9⅕　字数：280 千字
2008 年 9 月第一版　2008 年 9 月第一次印刷
印数：1—3000 册　定价：**39.00** 元
ISBN 978-7-112-09897-2
　　　（16705）
版权所有　翻印必究
如有印装质量问题，可寄本社退换
（邮政编码100037）

给鸿经、元中及佳欣

To HungChing, Dexter and Virginia

Preface

To complement theories developed in cognitive science that yielded some understanding of human intelligence, cross disciplinary research has begun to explore the intellectual aspects of creativity and the design knowledge created during the creative processes. Following this new trend, this book systematically explains well developed theories in cognitive science, their applications, and future possible directions of utilizing cognitive science in design. From explaining the fundamental concepts of cognition, elaborating advanced research methodologies, and discussing individual design thinking and creative processes through psychological experiments; phenomenon of design cognition should be understood better. Other studies, which embrace human cognition, including artificial intelligence and neuroscience are also covered briefly in this book to catch up.

Metaphorically speaking, the human mind is a black box, which has attracted many scholars to study the mental phenomena of thinking. Plato (427 ~ 347 BC) was the first to attempt to describe and categorize knowledge in writing. After the Renaissance, studies in human intelligence began moving from a philosophical approach to the perspective of exploring mental phenomenon. Research methods focused on exploring how knowledge is learned and formulated through observations; whereas in modern psychology, studies conducted through psychological experiments moved even further to define memory structure, recall methods, nature of linguistic and cognitive phenomena. Therefore, approaches and methodologies of study had changed gradually since 1930s.

After the digital computer was successfully developed, a new area of artificial intelligence that represented the mind using symbols was established. Following developments in the use of symbolic logic to represent knowledge in computers, the use of computation to simulate human intelligence fostered the study of human cognition. Cognitive science came into being; which studies the nature of mental tasks and the mental processes that perform them. Currently, various disciplines of computer science, linguistics, cognitive psychology, cognitive neuroscience, and philosophy are also involved in the study of the fundamental of intelligence and cognition.

In design, the application of cognitive science to study creativity is a novel paradigm that is waiting to be further developed for exploring design thinking. This book explains the concept of human cognition, its application in design, environmental information impact to thinking, as well as potential areas for future study. The basic conceptual framework is that design is a mental process involving many cognitive operations such as (1) problem solving; (2) retrieving stored information from memory for decision making; and (3) sequences of utilizing information during the design processes. These processes have many cognitive mechanisms involved. It is because of the variation of the mechanisms that different intellectual results are yielded.

Chapter One briefly explains the historical development of cognitive science, with mention of its roots in psychology, and progresses from cognitive psychology to the current status.

Methods for how computer intelligence can simulate human intelligence are explained. The major cognitive mechanisms utilized while the mind is processing information, theories applied in the field, and operational methodologies defined by scholars are also briefly discussed. Chapter Two introduces the correlations between

前 言

人类智慧已由认知科学中得到许多深入了解,在新科技飞速的前进演变中,跨领域学科研究也开始探讨在设计创造过程中所生成及运用的设计智慧。顺着这新趋势,本书条理分明地解释认知科学中已发展成形的理论、在设计中的运用,以及未来可能的走向。借着详述基本的认知观念,阐述深一层的研究方法,并经由心理实验来剖析设计思考和创造过程;相信设计的认知现象应可被充分了解。其他新兴的学科,以及或明或暗地与认知有关的学科,包括人工智能和脑神经科学等都会于本书中简略涵盖,以顺应新的未来科技发展。

隐喻而言,人类的脑海是个黑盒子,也因此吸引了无数学者研究思考的心智现象。西方哲学家柏拉图(公元前427~前347年)是西方史上第一个试着讨论智能,而且化成文字记录下来的学者。文艺复兴之后,人类智慧的研究就开始由哲学的探讨逐渐转换到人文心理现象的研究角度。研究的方法是经由观察去探讨智慧是如何学到和成形的。但在现代心理学里,研究方式则经由心理实验的运作,进一步探寻记忆的结构、追忆的方法、语言的本质和认知现象等。于是,研究的方法和方向从20世纪30年代以后就逐渐有了些变化。

在电子计算机发展成功之后,一些先驱立刻成功地发展出以象征符号来代表知识。于是,人工智能也因而成立。随着使用象征符号于计算机中仿真知识,并运作智能程序,计算机运算也就被用来模拟人类智慧,于是人类的认知心理现象就开始被有效地开放研究了。也因此,认知科学也就成功地形成一个新的研究领域。其着重研究发掘心智运作和探讨实现这些运作的心理过程。目前,这一学科已包括来自计算机科学、语言科学、认知心理学、认知脑神经科学和哲学等不同学科专家,共同研究智慧与认知的本源。

在设计中,应用认知科学去研究创造力还是一个新奇的范畴,有待更进一步地开发探讨设计思考。本书将说明认知科学的观念,在设计中的运用,环境信息会对思考产生什么影响,以及未来研究方向等。整个基本论证构架是把设计当作心智运作过程,在过程中有如下认知运作涉及:(1)解决设计问题;(2)在记忆中寻找信息作设计决策的搜寻过程;(3)在设计过程中运作信息的程序。这些过程都有许多心智制构参与运作。也因为这些制构的变化,而产生不同的智慧效果。

本书第1章介绍认知科学的历史变迁过程,由心理学的根源进化到认知心理学以迄今。在认知科学成为一门学科之后,研究方向开始着重于探讨如

cognition and design, especially the development of Gestalt psychology and its application in design, the historical changes of studying design theory from diagrammatic approach to the concern of human design thinking processes. Chapter Three explains the major components of cognitive mechanisms and cognitive operators that directly connect to cognition in the psychological processes.

Chapter Four explains the operational processes of mental images in design's related tasks. The unique knowledge representation of image, different from representation of verbal, is described and illustrated by a series of psychological experiments. Studies covered in this chapter explore the nature of mental imagery by exploring how images are used in the field of design. Two research models are developed. The first model proposes that architectural knowledge is organized in a hierarchical network in which elements are linked by architectural functions. The second model deals with the process of retrieving information from long-term memory. The process is divided into five stages, each involving a specific type of mental operation. These operations can be detected through reaction times. By using the framework of these two models, expert and novice designers are compared to observe the essences of architectural knowledge representation and mental image in design. Examples given in this chapter provide rigorous methods for conducting advanced psychological research in design. Parts of this chapter were originally printed in the Journal of Architectural and Planning Research, Volume 14, Number 1, pages 52~77. It is reproduced with permission. Chapter Five introduces a method to understand design from a problem solving point of view. A protocol analysis method is used on segmenting the mental process through a psychological experiment for analyzing design cognition. Parts of Chapter Five were printed in the Design Studies, Volume 11, Number 2, pages 60~80, Copyright 1990. It is reproduced with permission from Elsevier. Chapter Six explains how environmental information impacts human cognition, how virtual reality can be applied to test cognition virtually, and future possible directions in utilizing neuroscience to study design cognition. Parts of the chapter have been printed in the Proceedings of the 12th International CAAD Futures Conference, pages 373~386. Concepts had been used with permission from Springer.

Written in both English and Chinese, the context, meaning, and concepts included in this book are the same. Yet, the two versions are not simply results of merely word-to-word verbatim translation. In fact, in describing domain-specific knowledge, certain theoretical notions could not be precisely explained by equivalent words in different languages with different culture backgrounds. Thus, methods used in this book were partially direct translation with adequate words and modified syntax to illustrate the idea clearly. However, theoretical concepts needed to first be written in one language than translated into a second while preserving the meaning. The most difficult part was revision processes. Whenever a language is changed the other one must be changed simultaneously to maintain the meaning which doubled the author's work load. Hopefully, this book provides readers with enough theoretical background to capably conduct cognitive studies in design.

It is hoped, this book will be of interest to readers from the fields of design and engineering, to increase their understanding of human thinking in design and cognition in a living environment. Designers may better understand how cognition affects the design processes. By understanding the cognitive processes involved in the phases of a design task and the factors that influence the processes and outcomes, designers and engineers can modify design methods to improve design products. And finally, this book can provide knowledge to make the design process scientifically controllable for generating qualified products, which should have a broad appeal to those with interest within and beyond architecture.

何以计算机智能仿真人类智能的方法。至于人脑运作信息时,有哪些认知组构负责参与,有哪些理论已经发展出,以及有哪些运作方法已经定形等都在本章简短涵盖。第2章介绍认知和设计的关连,特别是完形心理学的发展和在设计中的运用,以及研究设计理论由图解的方式转移到考虑人类的思考过程之历史变化。第3章说明认知的主要组构要素,尤其是心理过程中与认知有绝对关系的认知运作单元。

第4章解释与设计有关的心智影像之运作过程。有异于语意知识,影像具有独特的知识表征。这表征也经由系列的心理实验于本章中详细地图解阐明。本章中的研究,经由分辨影像是如何用于设计领域中来探讨心智影像的本质。两个研究模式被发展出。第一个模式提出建筑知识可被看成是一个体系网络,由建筑机能联结而成。第二个模式解说在长期记忆中,信息是如何被撷取的。撷取过程可分为五个阶段,每一阶段都有特别的心智运作,这些运作可由心理的反应时间测出。通过运用这两个模式比较外行和专家设计师的设计,则建筑知识表征和心智影像用于设计的本质即可观察出。本章中所用的例子提供了范例和严谨的方法可供高深的设计心理研究。部分结果刊在《建筑计划及研究》季刊,第14卷,第1期,52~77页,允许再引用。第5章介绍由解决问题的角度来了解设计。本章中引用一心理实验例子以"原案口语分析法"解析心理过程,分析设计中的认知。部分结果刊在《设计研究》季刊,第11卷,第2期,60~80页,1990年版,也允许再引用。第6章述说环境信息对认知的影响,如何运用虚拟实境作虚拟心理研究,以及未来使用脑神经学研究设计认知的可能走向等。部分结果刊在《第12届计算机辅助建筑设计2007年国际研讨会专辑》,373~386页,观念也允许再引用。

本书以中、英文写成,其中观念、意义及文脉是相同的。但是两文之间不是单纯的字对字及一对一的直译结果。事实上,在描述专业性知识中,一些理论观念无法以两种语言中相似的字去正确描述,因为两种语言有不同的文化背景。因此本书所用的方法是直译再加上适当用字和贴切语法去清晰地解说一些理论观点。也因为如此,本书所花的精力是一般著作的两倍。基于过程是用一种语言去写出一些理论,然后把原文翻译成另一种语言,在这过程中还得考虑、配合原来的构思,此外在修改审稿时要兼顾构思和两个版本的语义说法。但愿耕耘的结果能提供读者以充分的理论背景,以有能力进行与认知相关的设计研究。

本书希望能吸引来自设计及工程专业的读者增加对人类设计思考过程,以及在居住环境中认知能力的了解。也愿此书能让设计师更了解认知对设计过程会有什么影响。充分了解设计过程中的认知程序之后,设计者将会更清楚何种因素会影响认知程序及结果,因而有效地修改设计方法以得到更好的设计成品。总而言之,本书将提供知识资源让设计过程能被科学系统化地操制,以得到更好的设计结果,而且这些介绍会让建筑界内及建筑界以外有兴趣的读者一起分享。

Acknowledgement

Experiments covered in Chapters 4 and 5 utilized seven subjects, including students, architects, and faculty members. The protocol analysis methods used in Chapter 5 were originally developed in the 1970's by Simon and Newell. It has been modified slightly in this research project for studying architectural design processes, which had been evaluated and tested together with Simon (1979 Nobel Economic Price laureate), who was one of the author's Ph. D. mentors at Carnegie Mellon University. Experiments covered in Chapter 6 are cooperative works carried out by the four members of the Architecture Cognition Group at Iowa State University, using 31 subjects. The author wishes to thank Omer Akin for his advice while the author was studying in Pittsburg, Linda Webster for providing editing assistances, the patience of editor Ms. Chen Hua for waiting for the completion of the book, and my family HungChing, Dexter, and Virginia, for their patience during the long writing processes.

感 言

第 4 章及第 5 章的实验有 7 位受测者，包括学生、建筑师和教授参与。第 5 章使用的原案口语分析法原先由司马贺及纽韦尔二人于 1970 年发展成熟。但在本研究案中作适度修改以便适合运用于建筑设计研究。修改时，曾和司马贺（获得 1979 年诺贝尔经济奖）一起测试过，因他是著者于卡内基梅隆大学修读博士学位时的导师。第 6 章中的实验是艾奥瓦州立大学建筑认知组四位组员的合作成果，有 31 位受测者参与。著者还要感谢奥码埃肯在匹兹堡时代的指导，琳达为本书所提供的文稿选订，编辑陈桦耐心等待本书的完稿，和家人鸿经、元中、佳欣一起度过这漫长的写作过程。

Contents

Chapter 1　Historical background and perspectives of cognitive science 2
　1. 1　Original studies of human knowledge in philosophy 2
　1. 2　The origin of psychology 6
　1. 3　The establishment and development of psychology 6
　1. 4　The formation of cognitive psychology 10
　1. 5　From cognitive psychology to cognitive science 12

Chapter 2　Correlations between design and cognition 20
　2. 1　Design principles and methods 22
　2. 2　Gestalt psychology and design 22
　2. 3　Design methodology 30
　2. 4　Cognitive science and design thinking processes 34

Chapter 3　Cognitive mechanisms 38
　3. 1　Information processing theory 38
　3. 2　Perception and attention 40
　3. 3　Pattern recognition 46
　3. 4　Mental image 50
　3. 5　Memory 52
　3. 6　Mnemonics 58

Chapter 4　Mental image and cognitive processing 62
　4. 1　Mental imagery in design 62
　4. 2　Imagery in memory 64
　4. 3　Mental imagery representation of dual codes 68
　4. 4　Example of mental processing of imageries 70
　4. 5　Mental Images and Design 92

目 录

第1章 认知科学的历史背景与综观 …………………………………………… 1
  1.1 哲学中对智慧的探讨 ……………………………………………………… 1
  1.2 心理学的起源 ……………………………………………………………… 5
  1.3 心理学的成立及发展 ……………………………………………………… 7
  1.4 认知心理学的成立 ………………………………………………………… 11
  1.5 认知心理学到认知科学 …………………………………………………… 13

第2章 设计与认知的关联 ……………………………………………………… 19
  2.1 设计原则及方法 …………………………………………………………… 21
  2.2 完形心理学与设计 ………………………………………………………… 21
  2.3 设计方法论 ………………………………………………………………… 29
  2.4 认知科学与设计思考过程 ………………………………………………… 33

第3章 认知的组构要素 ………………………………………………………… 37
  3.1 信息处理理论 ……………………………………………………………… 37
  3.2 知觉与注意力 ……………………………………………………………… 39
  3.3 形态辨认 …………………………………………………………………… 43
  3.4 心智影像 …………………………………………………………………… 47
  3.5 记忆 ………………………………………………………………………… 51
  3.6 记忆增进术 ………………………………………………………………… 57

第4章 心智影像与认知运作过程 ……………………………………………… 61
  4.1 与设计有关的心智影像 …………………………………………………… 61
  4.2 记忆中的心智影像 ………………………………………………………… 63
  4.3 记忆中影像呈现的双码论 ………………………………………………… 65
  4.4 心智进行影像处理的实例 ………………………………………………… 69
  4.5 心智影像与设计 …………………………………………………………… 93

第5章 设计中的知识运作 ……………………………………………………… 97
  5.1 问题解决理论 ……………………………………………………………… 97

Chapter 5   Knowledge operations in design 98
   5.1   Problem solving theory 98
   5.2   Concepts of Problem solving theory 100
   5.3   Representation and search strategies 104
   5.4   Concepts of protocol analysis 106
   5.5   Methods of collecting protocol data, coding, and analysis 108
   5.6   Applications of protocol analysis 134
   5.7   Advantages and disadvantages of protocol analysis 136
   5.8   Other methods used for data collections on thinking 138

Chapter 6   Impacts of technology to cognition 142
   6.1   Cognitive science and artificial intelligence 142
   6.2   Cognitive theory in design 144
   6.3   Cognition in virtual reality, the sense of presence 146
   6.4   Neuroscience and neurocognition 152
   6.5   Conclusions 154

Bibliography 157

Chinese – English Translation Index 167

  5.2 问题解决理论的概念 …………………………………………………… 99
  5.3 表征呈现和搜寻策略 …………………………………………………… 101
  5.4 原案口语分析概念 ……………………………………………………… 103
  5.5 原案口语数据收集、编码及分析法 …………………………………… 105
  5.6 原案口语分析的运用 …………………………………………………… 133
  5.7 口语分析的优点及缺点 ………………………………………………… 135
  5.8 收集思考数据的其他方法 ……………………………………………… 137

第 6 章 科技对认知的影响 ……………………………………………………… 141
  6.1 认知科学与人工智能 …………………………………………………… 141
  6.2 认知理论与设计 ………………………………………………………… 143
  6.3 虚拟空间中的认知,身临其境的存在感 ……………………………… 143
  6.4 网络神经和网络心理学 ………………………………………………… 151
  6.5 总结 ……………………………………………………………………… 155

参考书目 …………………………………………………………………………………… 157
中英对照检索 ……………………………………………………………………………… 167

第1章 认知科学的历史背景与综观

人是最独特的生物,因为人类会学习、会思考、会推理、会组织见解,于是知识就成形,智能就存在了。也因为人类具有智慧,智慧就成了一种驱动力,造就了科技文明,也持续地把文明往前推进。这一切都要归功于智慧的存在。也因为智能的价值被肯定,吸引了无数不同领域的学者去发掘它的自然本质。因此,在人类的智慧史里,自古以来,已有不少学者在不同领域中探讨下列基本的问题:(1)知识与智能来自何处?(2)知识如何呈现在人脑里?(3)知识包含些什么?

在探讨这些问题的过程中,有不少学科涉及,也有不少理论被哲学、心理学、认知心理学、认知科学、神经心理学等领域发展出并建议过。但与人类智慧有关且研究最深入的一科,应归功于认知科学。认知被定义成"知识的获取"。而认知科学这领域合并了认知心理学和计算机科学,用计算机仿真人脑如何吸入、呈现、处理及运用知识。因此,产生许多适切的科学化解释。本章首先简略地重点介绍知识及智能被探讨过的历史过程,和认知科学如何成为一个杰出研究领域的来龙去脉。

1.1 哲学中对智慧的探讨

最早研究知识和智能的学者是哲学家。早期的哲学偏重于分析宇宙、自然、人性、宗教及逻辑。三位有名的希腊学者,苏格拉底(公元前469~前399;图1-1及附记1)、柏拉图(公元前427~前347;图1-1及附记2)、亚里士多德(公元前384~前322;图1-1及附记3)最早发展出一套经由通过

图 1-1 苏格拉底(公元前469~前399)、柏拉图(公元前427~前347)、亚里士多德(公元前384~前322)
Figure 1-1 Socrates (469~399 B.C.), Plato (427~347 B.C.), and Aristotle (384~322 B.C.)

Chapter 1 Historical background and perspectives of cognitive science

Humans are unique beings because of their ability to learn, think, reason, and organize their thoughts; thus knowledge is shaped and intelligence formed. The intelligence owned by human beings is what makes technology and civilization possible, and is the unique force that keeps the universe revolving. Recognizing the value of intelligence, scholars in various fields have long studied it to discover its essential nature. In human intellectual history, scholars from various disciplines have studied the basic questions of: (1) where do knowledge and intelligence come from, (2) how are they represented in the human mind, and (3) of what do they consist?

In exploring these questions, a number of theories have been developed and proposed by fields of philosophy, psychology, cognitive psychology, cognitive science, and neuropsychology. One of the most insightful studies in human intelligence has been cognitive science. Cognition is defined as the acquisition of knowledge. Cognitive science combines cognitive psychology and computer science, applying computers to simulate how human brains acquire, represent, process, and utilize knowledge. This has yielded many appropriate scientific explanations.

This chapter will first explain the important historical developments in the study of human knowledge and intelligence, and describe how cognitive science has become a promising research field.

1.1 Original studies of human knowledge in philosophy

The earliest scholars studying knowledge and intelligence were philosophers. Early philosophy concentrated on analyzing the nature of the universe, the nature of the human mind, logic, and religion. Three Greek philosophers, Socrates (469~399 B. C. ; Figure 1-1 and note 1), Plato (427~347 B. C. ; Figure 1-1 and note 2), and Aristotle (384~322 B. C. ; Figure 1-1 and note 3), were the first to develop methods for posting a series of questions and answers, and conducting cases to obtain empirical observations on truth, which shaped the basic branches of philosophy——metaphysics (concerning the cause-and-effect relationships between things in the world), epistemology (concerning the reality of truth), ethics (concerning the meaning of life and the conduct of human moral relations), and logic (explores

询问系列问题的方法，提供个案，进行例证，观察真理，而成立了几门哲学内的基础学科——包括形而上学（关切世上所有事情的前因后果关系）、知识论（关切如何认识真理）、伦理学（关心生命意义与道德实践）以及逻辑（探讨达到正确思考结果的规律和方法）。其中，逻辑这一派专注于人类如何思考推理。亚里士多德的三段逻辑论就试着去解说知识的自然法则。目的是让人成为一个批判思考者，也由逻辑训练出更犀利的思考能力，作出正确的考虑。罗马时期的哲学家也沿袭希腊哲学学说，继续探讨实践的法则。

基督教时期（公元400~1400）的哲学观围绕在基督教的人、神，与世界的关系。圣·奥古斯丁（354~430；图1-2及附记4）在公元400年写的《神之都市》，从神学和目的论的角度去诠述人、神、宇宙与真理的关系，是中古时期最重要的哲学著作。但中古世纪哲学认为真理是由神中获得的想法只持续到公元1400年左右。而后，哲学开始脱离神学而更相信理性。随此，文艺复兴时代兴起（1300~1600）。文艺复兴期间，人们开始对与他们有关的世界感兴趣。哲学就把注意力由超自然转向自然，专注于地球上发生的经历。这时期的哲学相信真理是由理性中寻求，知识是由演绎推理得来。其基本方法是由一个主要前提中诱导出一个学说理论，或由数个已知事实中得出新的结论。这段时期中的科学家都成功地发展出一些科学研究方法。例如哥白尼（1473~1543；图1-2及附记5）就用数学解释出地球和其他星球是绕太阳公转的。伽利略（1564~1642；图1-3及附记6）是第一位用望远镜观察星球的科学家，而且也证明不同重量的物体降落时的速度是相同的。至于开普勒（1571~1630；图1-3及附记7）则提出了行星运动三大定律等。

图1-3 伽利略（1564~1642）与开普勒（1571~1630）
Figure 1-3 Galileo Galilei（1564~1642）, and Johannes Kepler（1571~1630）

到了1700年，知识论已变得日益重要。哲学的疑虑围绕于人是如何得到知识和如何知道真理的问题上。他们开始相信"知识的模式"是环绕在机械及物理的领域中。哲学家认为实地验证是得到知识的法门，因为经过操作观察和经验体会，根本的基础概念就会衍生出来。牛顿（1642~1727；图1-4及

图1-2 奥古斯丁（354~430）与哥白尼（1473~1543）
Figure 1-2 St. Augustine（354~430）, and Nicolaus Copernicus（1473~1543）

图1-4 牛顿（公元1642~1727）
Figure 1-4 Isaac Newton（1642~1727）

the rules and methodologies for achieving precise thinking results). The branch of logic focused on how humans think critically. Whereas syllogistic logic by Aristotle tried to explain the nature of rules used in formulating knowledge. The purpose was to make a person a critical thinker by sharpening the logic abilities to think precisely. Roman philosophers followed Greek lines of study on thought and worked on rules of conduct.

The Christian era (400~1400) of philosophy centered around the relationship between God, man, and the world. Saint Augustine's (Aurelius Augustinus, 354~430; Figure 1-2 and note 4) "*City of God*" written approximately 400 AD, elaborated on the relationships between God, man, universe, and truth from theological and teleological points of view. It was one of the most important philosophic works of the Middle Ages. The notion of that truth was believed to come directly from God in the Christian era lasted until approximately 1400. Later, the philosophy came to believe more on reason and separated from theology. Following this, the Period of Renaissance (1300~1600) came. In this period, men became fascinated with the world about them. Philosophy turned its attention from super natural to nature, concentrating on experiences occurring on earth. Truth could be found through reason. Knowledge could be obtained from deductive reasoning. The basic method was to draw a theory from a major premise, or reach a new conclusion from known facts. Scientists in this period developed successful methods of investigation. For example, Nicolaus Copernicus (1473~1543; Figure 1-2 and note 5) explained through mathematics that Earth and the other planets revolve around the sun. Galileo Galilei (1564~1642; Figure 1-3 and note 6) applied the telescope to observe stars, and also demonstrated that different weights descend at the same rate. Johannes Kepler (1571~1630; Figure 1-3 and note 7) postulated the three fundamental laws of planetary motion, etc.

Approximately 1700, epistemology became important. Studies in philosophy speculated on the questions of how men acquired knowledge, and knew truth. They theorized that models for knowledge involved mechanics and physics. Philosophers used empirical approaches and believed that knowledge could be built upon observation and experiments, which would give rise to fundamental ideas. Isaac Newton (1642~1727; Figure 1-4 and note 8) combined theories from Copernicus, Galileo, and Kepler to describe the world as a giant machine through mathematics. Through empirical observation, he discovered gravity.

Philosophers of the 1800's turned their attention to various aspects of human experiences, and the human being became the focus of philosophic attention. Immanuel Kant (1724~1804; Figure 1-5 and note 9) stated that all ideas are representations of sensory experience. Specifically, our impressions come from our senses, but it is our minds that shape and organize these impressions to make them meaningful. The mind accomplishes knowledge through a priori (that knowledge generation is from inducing causes to effects) or rational judgments that do not depend on experiences (Kant, 1998). Since the 1900s, this philosophy has moved in two major directions. One follows the development of logic, mathematics, and science; the other, the increasing concern about man himself. Bertrand Russell (1872~1970; Figure 1-5 and note 10) turned his focus onto the philosophy of science, especially on man's ability to know and to use scientific methods. There were other philosophic concerns regarding man's survival in, and adjustment to, this

附记8）就使用数学来形容这世界是个大机器，并且结合哥白尼、伽利略和开普勒的学说，经过实地观察，进而发现地心引力。

1800年代的哲学家开始把哲学注意力转移到人类多方面的不同经验上。人变成哲学的研究焦点。康德（1724~1804；图1-5及附记9）就表明了所有人的概念是感官经验的表征呈现，人对事物的印象是经由五官而得到的。但也因为是人的心智促成这些印象并组织这些印象，所以印象才变得有意义。人的心智运作生产出知识意义的活动，是经由（从原因推理到结果的）先验推理法或条理判断方式来产生意义，而不是单纯由经验体会而得来的（Kant, 1998）。自20世纪以来，哲学移到两个主要大方向：一是沿袭逻辑、数学和科学的发展；另一是着重关切人本身。例如罗素（1872~1970；图1-5及附记10）就将他的重点转向科学哲学上，特别专注在人对科学的领悟和运用科学方法的能力上。其他的哲学关切则围绕着人如何在这个多变化的世界中去调整自己让自己生存。总而言之，这些简短历史介绍解释了哲学学派在探讨人类知识和智能的路程变化。

1.2 心理学的起源

在1800年以前，心理学即已是哲学的一个支派。心理学（Psychology）这个词由希腊文"psyche"（心、灵或精神）及"logos"（研究或学科）演变而来。按照字面上的意义解释，心理学是"研究心灵或精神"的学科。在古希腊，柏拉图、亚里士多德与一些古哲学家就有理论曾经解释过人类行为的缘由，和心智与人身的关系。他们相信心智是一个可辨认的个体，位在人身体上某一特定区域。在1600年代到1700年代，一些哲学家相信心智可被划分成几部分天赋才能，包括意愿、思考、推理等。这些才能是促成不同行为的缘由。另一些哲学家则相信心智在出生时是空的，人必须要经过经历才会发展出概念。例如法国笛卡儿（1596~1650；图1-6及附记11）相信心与身是不同但互连的。这心影响身，身影响心的互联二元论反映出现代心理学中"生物学"的倾向。洛克（1632~1704；图1-6及附记12）相信心智历程完全是学自于经验。这经验论想法影响了现代科学方法和学习理论。在1800年代中期，科学家开始严谨地研究人类动机、思考、感觉和情

图1-5 康德（1724~1804）与罗素（1872~1970，数据来自罗素档案室、麦克马斯特大学图书馆）
Figure 1-5 Immanuel Kant（1724~1804）and Bertrand Russell（1872~1970, resource of Bertrand Russell archives, McMaster University Library）

changing world. In summary, this brief historical introduction explains the developmental changes of philosophic studies on human knowledge and intelligence.

1.2 The origin of psychology

Before 1800, psychology was simply an area of philosophy. The word psychology comes from the Greek root "psyche" (mind, soul, or spirit) and "logos" (study or discussion). Literally, psychology is the study of the human mind or soul. In ancient Greece, Plato, Aristotle, and other ancient philosophers developed theories about the causes of behavior and the relationship between mind and body. They believed that the mind, located in a specific part of the body, was an identifiable entity. In the 1600's and 1700's, some philosophers believed that the mind could be divided into several inborn facilities, including will, thought, and reason. These faculties accounted for the causes of different behaviors. Others believed that the mind is empty at birth and must have experiences to develop ideas. For example, Rene Descartes (1596~1650; Figure 1-6 and note 11), believed that mind and body are different, but interrelated. This interrelated dualism, that the mind influences the body and the body influences the mind, has been reflected in the biological approach in modern psychology. John Locke (1632~1704, Figure 1-6 and note 12) believed that mental processes are learned entirely through experience; his empiricism influenced specific methodologies and learning theories. By the mid-1800's, scientists had begun to develop rigorous studies on human motivation, thought, feeling, and emotion. This range of study is the area covered in the field of psychology today.

In short, prior to the 18th century, psychologists believed that the body and mind were two distinct entities. Studies in psychology were limited to the level of the mind and soul, methods applied were subjective and lacked empirical evidences. In the 19th century, studies became systematic, objective, and relied on scientific experiments conducted in laboratories. Psychologists began to believe in the unity of body and mind, and realized that living beings responding to stimuli have invisible external reactions in behavior, action, or physiological changes which are measurable; and invisible internal reactions described variously as motive, sensation, thought, and feeling. For example, Gustav Fechner (1801~1887; Figure 1-7 and note 13) explored the relationship between the physical characteristics of a stimulus and the psychological experience of that stimulus. Herman von Helmholtz (1821~1894; Figure 1-7 and note 14), a German pioneer in experimental psychology, studied reaction time, speed of nerve conduction, the workings of the eye, and the color vision process. Their research contributed to the establishment of psychology as a scientific discipline. Thus, psychology, gradually separated from philosophy, became an independent science. Modern psychology applies systematic methodologies to measure, judge, predict, and evaluate the visible and invisible relations in human intelligence.

1.3 The establishment and development of psychology

Psychology, recognized as a science based on careful experimentation and observation, is generally acknowledged to have been established in 1879. In 1874, Wilhelm Wundt (1832~1920, Figure 1-8 and note 15), the father of psychology, published "*Principles of Physiological Psychology,*" which is recognized as one of the most important publications in the history of the field. In 1879, he created one of

感。这些范畴就是今日心理学涵盖的领域。

图1-6 笛卡儿（1596~1650）与洛克（1632~1704）
Figure 1-6 Rene Descartes (1596~1650) and John Locke (1632~1704)

简而言之，18世纪以前，心理学家认为身和心是分离二次元。一般心理学的研究也局限于心灵层次，方法主观、缺乏科学实证的学科。但从19世纪开始，研究转向系统化、客观化，并依赖在实验室中做的科学性实验研究。心理学家开始相信人类的身和心是一体的，也了解到生物对外界的反应有一种是外显可测的行为、动作或生理变化，和另一种内隐而难测的内在反应，如动机、情绪、思考和知觉等。例如，费希纳（1801~1887；图1-7及附记13），一位德国实验心理学的先驱，就曾发掘诱因的物理特性和该诱因与心理经验的关系。赫姆霍兹（1821~1894；图1-7及附记14）研究反应时间、神经发动的速度、眼球机能和视觉色彩过程。这些研究的贡献是把心理学变成一门科学学科。因此，心理学才从哲学中脱离出来，成为一个独立科学。现代心理学则用系统化的方法，由内隐及外显的反应关连对人类智慧作测量、判断、预测以及评估。

1.3 心理学的成立及发展

心理学，是一门被承认是基于仔细实验、用心观察的科学，被公认是在1879年建立的。1874年，冯德（1832~1920；图1-8及附记15）这位心理学之父，出版了《生理心理学的原则》，这在心理学史上被认为是重要的著作之一。1879年，他在德国莱比锡创立了一个领先的心理实验室，研究医学和生理心理机能（詹姆斯在美国于同年创立第二个心理实验室）。冯德的概念是用**内省法**去形容并分析意识经验，包括心情知感、影像和感觉。冯德及其学生，被分类为**结构主义心理学派**，认为意识经验是可以由成分组构而成的。所以，此派主要研究意识的结构和组成元素。使用的内省法是让训练过的观察者，在被细心控制的情形里，通过口语报告他们内心意识到的经验内容，并分析该经验内容。这方法可得到一些证据说明内心操作的形态。但不同的实验室有各自特殊的学说理论，运作出的内省报告就会有不同的形式结果。这说明了内省法无法提供一个透明的窗口去观察内心的运作。

詹姆斯（1842~1910；图1-9及附记16），是一位非常有影响力的美国心理学家。不同于结构主义心理学派，他的理论强调使用实际方法研究人对环境的适应能力。他的研究，被划分为**机能主义心理学派**，启蒙了教育心理学和工业心理学两个领域。1913年，美国心理学家华生（1878~1958；图1-9及附记17）引进**行为主义心理学**。华生认为，心理学家应研究、观察人类和动物可观、可测的外显行为，而非经验。行为主义学派反对结构主义学派的内省法，强力主张只有分析人类行为和反应，才是

图1-7 费希纳（1801~1887）与赫姆霍兹（1821~1894）
Figure 1-7 Gustav Fechner (1801~1887) and Herman von Helmholtz (1821~1894)

the first two psychological laboratories in Leipzig to study medicine and physiology (William James established the second one in America the same year). Wundt's concepts used **introspection** to describe and analyze conscious experiences including sensations, images, and feelings. Wundt and his students, called "**structuralists**," thought that conscious experience could be constructed through components. They concentrated on exploring the structural components of conscious experience. This method of introspection required highly trained observers reporting the contents of their consciousness under carefully controlled conditions, and analyzing the reported experiences. This method obtained some evidence illustrating patterns in the working of the mind. But different laboratories were reporting different types of introspections, which emanated from the particular theory generated by the laboratories. Thus, the introspection method did not provide a transparent window to view the workings of the mind.

William James (1842~1910, Figure 1-9 and note 16), an influential American psychologist whose theories differed from structuralism, emphasized a practical approach regarding adaptability of people to their environments. His emphasis, termed **functionalism**, led to the development of educational and industrial psychology. In 1913, American psychologist John Watson (1878~1958; Figure 1-9 and note 17), postulated **behaviorism**, proposing that psychologists should study the observable, measurable behavior of men and animals, not their experiences. Behaviorists, rejecting methods of introspection used by structuralisms, believed that the analysis of behaviors and reactions was the only objective method to get insight on human reactions. Watson discovered that animals could be trained to respond to certain stimuli by **association**. This behaviorism group conducted laboratory studies on **stimulus-response** relationships in behavior, measuring how people responded to given stimuli. Many psychologists agreed that experimental methods to study behavior were most valid. But some thought that behaviorism neglected thought processes and the development of human personality. By 1920, behaviorism replaced structuralism as the dominant study in the field of psychology. Watson's behaviorism had influenced psychologists' work in learning theory.

Gestalt psychology was founded in Germany around 1912 by Max Wertheimer (1880~1943; Figure 1-10 and note 18) and his associates Wolfgang Kholer (1887~1965, Figure 1-10 and note 19) and Kurt Koffka (1886~1941; Figure 1-10 and note 20). The name "gestalt" means shape, form, special pattern, and configuration in the physical form. Gestalt psychology concentrated on analyzing how human beings obtain visual perception from visual stimulus. However, analyses addressed perceiving the whole observed entity rather than the individual parts. Gestalt psychologists disagreed with the structuralists' methods of observing self descriptions to analyze components of sensual experiences, and their theories that experience could be broken down into parts of hearing, seeing, and feeling. They also disagreed with the behaviorists over their emphasis on animal experimentation and disregard for the mental processes. On the contrary, Gestalt psychology believed that all factors should be studied as a whole in order to comprehend their relationships instead of perceiving individual unrelated parts; their basic principle being that the whole is greater than the sum of its parts, or the whole is not the sum of the parts. Further, human recognition of a visual image generates, through perception, an organized overall

图 1-8 冯德 (1832~1920)
Figure 1-8 Wilhelm Wundt (1832~1920)

图 1-9 詹姆斯 (1842~1910) 与华生 (1878~1958, 图片来自艾克隆大学心理学历史档案室)
Figure 1-9 William James (1842~1910) and John Watson (1878~1958, Psychology Archives–The University of Akron)

探索人类反应的惟一客观方法。他发现动物可以训练成对某种刺激经**联想**而作回应。行为主义心理学派，曾做了一些实验室的研究，分析外显行为上的**刺激和反应**间的关连；亦即实验者测量人对外在刺激物如何作反应。不少人同意心理学是应该依赖实验方法去研究行为的，但也有学者认为行为主义学派忽略了一些思考过程和人性发展问题的研究。不过行为主义学派从 1920 年开始就取代了结构主义学派，成为心理学的主流。而且华生的行为主义学派确实影响了一些心理学的理论。

完形心理学（或格式塔心理学）由韦特墨（1880~1943；图 1-10 及附记 18）和他的同事苛勒（1887~1965；图 1-10 及附记 19），及考夫卡（1886~1941；图 1-10 及附记 20）于 1912 年在德国创立。完形此名是中文意译，意指完整形态，其中文发音直译，则是格式塔，意指形状、造型、特殊形态和实体的配置。完形心理学重在分析人类如何由视觉刺激产生视觉上的认知。但分析注重在审视对象的视觉整体，而非单元部分。完形心理学派不同意结构心理学派的以观察自我描述法去分析意识经验成分的做法，以及经验是可以被分裂成听、看、感受的单元部分之理论。也不

图 1-10 韦特墨 (1880~1943)、苛勒 (1887~1965) 及考夫卡 (1886~1941)（图片来自艾克隆大学、心理学历史档案室）
Figure 1-10 Max Wertheimer (1880~1943), Wolfgang Kholer (1887~1967), and Kurt Koffka (1886~1941) Photos are from Psychology Archives—The University of Akron

shape, but not a combination of individual images. The relationships between parts of a given stimulus, when perceived as a whole, would give true meaning. Thus, Gestalt psychology has been recognized as the basis of cognitive psychology.

Gestalt psychology had obvious influence on concepts of perception, creativity, problem solving, and therapy. But, some concepts and hypotheses applied by the Gestalt psychologists were difficult to implement through empirical experiments. Thus, its limited application restricted continuous supports. This movement lasts for about 20 years. During the same period of Gestalt psychology, there was another movement, founded by Sigmund Freud (1856~1939; Figure 1-11 and note 21). This movement emphasizes the influence of unconscious forces, impulse, and internal conflict on everyday behavior. Around the same period, Jean Piaget (1896~1980; Figure 1-11 and note 22) proposed a series of theories on cognitive development (Piaget, 1967; 1985). Piaget, a Swiss developmental psychologist, theorized that people develop cognitive **schema** (knowledge modules). Schema is the representation in the mind of a set of growing perceptions, ideas, and actions, which form categories of knowledge. In the growing processes of human beings, external information, including social interaction and physical experience, would provide stimulus to the existing schema (module). The human mind would build upon existing schema adjusting it by either assimilating or accommodating new information. This process fits new information into existing knowledge modules (assimilation) and alters the existing knowledge module in order to accommodate new information (accommodation). With the gradual alternation of the schema structure (knowledge module), intelligence grows.

All these emerging movements and schools helped explain the dynamic perspectives involved in the study of the essential nature of human beings. Research could be approached from: (1) the biological perspective, to study the relationships of biological systems of the brain and nervous to behavior and mental processes; (2) the behavior perspective, to analyze the influences of the environment on an individual's behavior and process of learning; or (3) the psychodynamic perspective, to study drives, needs, emotions, purposes, and wishes. Currently, the major academic groups in psychology include: (1) **experimental psychology**: handling biological and psychological foundations of behavior with experimental methods, which exclude case study and interview methods; (2) **social psychology**: handling group interaction, i. e., how thoughts, feelings, and behavior are perceived, influenced, and related to others; and (3) **developmental psychology**: studying the progressive psychological changes from infant to ageing, including the processes of growth and development. These areas all address to some degree cognition, and relate, to some extent, to cognitive studies. But, in order to explore in detail human perception, memory, attention, and problem solving ability, one must focus primarily on cognition.

1.4 The formation of cognitive psychology

Cognitive psychology became part of psychological research between the late 1950s and early 1960s. It is difficult to pinpoint the exact beginning of development in this field, because cognitive psychologists contributed parts of its body of knowledge over a span of years. Early in 1890, William James published "*Principles of Psychology*," which included chapters on attention, memory, imagery, and reasoning.

同意行为主义心理学派过于强调动物实验，而忽略心智历程的做法。相反的，完形心理学家相信所有经验因素应由一个全局的角度来研究整体，以便了解因素间的关系，而非看无关的部分单元个体。基本原则是说整体要比单元的综合还要重要，或全部不等于部分的综合。进一步而言，因为人对任何视觉图像的认知是经过知觉系统组织后产生的综合新意象，而非所有独立部分的意象集合。因为当我们去审视一个给予物体之构件关系时，如由整体性来看它，视者才会得到元素间关系的真正意义。因此，完形心理学也被认为是认知心理学的起步。

完形心理学派对视觉、创造力、解决问题和治疗的一些观念想法有明显的影响。但有些完形心理学家的学说与假设，很难经由科学的实验加以验证。因此，有限的运用条件限制了它的后续支持。这学科持续了大约二十年。与完形心理学派同时，另有一支学派——心理分析派，由弗洛伊德（1856~1939；图1-11及附记21）创立。这学派专注于分析无意识的动力、冲动和发生于日常生活中的内心冲突。在同一时期，让·皮亚杰（1896~1980；图1-11及附记22），也提出一系列的认知发展理论（Piaget, 1967；1985）。皮亚杰是瑞士的发展心理学家，他认为人类会发展认知**基模**（即知识模式）。基模代表在人脑中逐渐成长的知识。这基模是成套的知觉、观念或动作知识存于脑海中而形成的知识团块。在人类的成长过程中，所有外界信息，包括人际交流、物理经验等，都会提供已存基模（组块）的一些刺激。人脑便会根据已有的基模去整合同化新信息；或者调适已存的基模以接受新信息。这过程可解释为一方面协调新信息以便纳入已存的基模中（同化），或修改已存的基模以配合外来的新信息（调适）。由于基模结构（知识模式）的逐步改变修正，智慧因而成长。

图1-11 弗洛伊德（1856~1939）及皮亚杰（1896~1980，图片来自艾克隆大学、心理学历史档案室）
Figure 1-11 Sigmund Freud (1856~1939), Jean Piaget (1896~1980, Psychology Archives-The University of Akron)

所有这些提及的新兴运动及学派也说明了研究基本人性所涉及的活跃角度。比方说研究可从生物角度来探讨脑神经系统和心智行为的关系，也可从行为角度去分析环境对人类学习行为的影响，更可从心理动态去解析驱动力、需求、情感、目的和希望等。目前，在心理学中主要的学术专业包括：（1）**实验心理学**：采用实验方法研究心理问题，专注于行为中的生物和心理基础，但实验心理学不认为个案研究和问卷调查是实验方法；（2）**社会心理学**：专注于群体互动，例如研究个人的思想、感觉和行为是如何被别人的知觉影响或关连到的；（3）**发展心理学**：研究人从出生到老化之间的心理进化及改变，专注于生命的成长和发展过程。这些专业都会部分涉及认知，也与某些认知研究有关连。但如要详细探讨个人知觉、记忆、注意力和解决问题的能力，则认知才是主要的焦点。

1.4 认知心理学的成立

认知心理学大约在1950年代晚期到1960年代初期之间，变成心理学学科研究的一部分。一个学科的成立，是很难在发展过程的时间表上明定这个

These components contributed to its study. In 1924, Watson's "*Behaviorism*" provided the main stream of research method on **stimulus–response (S-R)**. In addition, Wundt's introspection and Gestalt psychology explored mental processes and visual cognition. All these works were first initiatives in the field. Yet, major research in this field concentrated on mental processes and specific cognitive activities in attention, perception, problem solving, memory, and language. Ulric Neisser (see note 23) first used the term of "Cognitive Psychology" in his book title published in 1967. He referred to the field as "the study of all processes, by which the sensory input from the environment is transformed, reduced, elaborated, stored, recovered, and used." Three years later, the journal "*Cognitive Psychology*" was published, which gave focus to ongoing research. In retrospection, the formation of this field was influenced by several factors.

The major influential factor was the development of what was called **human information–processing approach**, which came from **information theory**. Information theory, popular during World War II, is a branch of communication sciences that provides a way of analyzing the processing of knowledge. The acquisition, storage, retrieval, and utilization of knowledge in human mind was seen as a number of separate stages, with the information processing approach used to identify what happens at each stage. In 1958, the British psychologist Donald Broadbent proposed one of the first research models (Broadbent, 1958). His model showed the relation between perception and auditory attention. This established information–processing as the dominant research approach in this field.

The second major influence came from the development of digital computers in 1937 at Iowa State University. The capacity of digital computers led to active work in artificial intelligence, which attempted to program computers to perform intelligent tasks (Miller, Galanter & Pribram; 1960) and to simulate human information–processing (Simon, 1957). This led to incorporating concepts from computer science in psychological studies. The third influence was from linguistics. Noam Chomsky (1957) developed a mode of analyzing the structure of language. He proposed that language should be learned from a system of grammatical rules. This assumed that language is based on complex mental processes that affect the structural correctness of sentences and the understanding of their meanings. This influence brought cognitive psychologists together to identify new ways of studying language. All these concepts developed over time with Neisser including them in a filed of study, named **cognitive psychology**.

1.5 From cognitive psychology to cognitive science

The fundamental aspect of cognition is human information processing that human beings transform, reduce, elaborate, store, retrieve, and utilize **information**. "Information", in this regard, is knowledge. As long as knowledge can be represented numerically and programmed, the processes can be simulated in computers which, in turn execute human intelligent tasks. The influences of artificial intelligence from computer science and the study of higher mental processes and structures led to the new interdisciplinary area of **cognitive science**. Cognitive science concentrates on the understanding of mental representations, the analyses of thinking, and computer models of human thinking (Gardner, 1985).

The following example explains the concept. A human mind can be seen analogously as a big machine that accepts information input and

学科正确起始点的，因为这领域的知识成形需由心理学家经过一段时间的贡献。早在 1890 年，詹姆斯出版的《心理学原则》一书就有数章解说注意力、记忆、意象和推理。这些项目是现代认知心理学的学习范围。在 1924 年，华生出版的《行为主义》就提供了**刺激—回应**的主要研究方法。另外，早期心理学中冯德的内省法以及完形心理学，都曾探讨内心历程和视觉认知特点。这些都算是这学科中最早的启蒙开创成果。不过认知心理学的重点比较专注于内心心智的历程和认知活动，如注意力、视觉、解答问题、记忆和语言等。特别要说明的是这个名词是奈瑟尔（见附记 23）在 1967 年出版的《认知心理学》的书名中定了下来的。他归类认知心理学涉及研究所有的知觉过程，这一过程是人把由环境中得到的感觉数据，作了转移、减缩、叙述、储存、回收和利用。三年后，《认知心理学季刊》创刊，也替这领域定了位。但回首沉思，这学科的成形受到几个因素的影响。

最主要的影响因素是已发展出的**人类信息处理法**。此法来自**信息论**。**信息论**，兴起于第二次世界大战，是信息传递科学的一个分支。该理论提供了分析处理知识过程的构想。人类知识的吸取、储存、回收和利用，可视为许多分离的过程步骤。而信息处理法即用来确认每个步骤中所发生的事件。1958 年，英国心理学家布劳德本特，以信息分析法提出了第一个研究模式（Broadbent, 1958）。该模式显示了知觉和音觉注意力之间的关系，也将信息处理法变成认知心理学科中主要的研究方法。

第二个影响来自 1937 年在美国艾奥瓦州立大学发展出来的电子计算机。电子计算机的能力带动了人工智能的兴起，米勒等人试图把计算机程序化在计算机里去执行有智能的项目（Mill, Galanter & Pribram, 1960），司马贺也提出了仿真人类信息处理法（1957，见附记 24）。这诱导出一个把计算机科学中的构想观念应用在心理研究的趋势。第三个影响来自语言学。曲姆斯基（1957）发展出一套分析语言结构的式样。他表示学习语言应学自一系列文法规则。这一观念说明语言本身是个复杂的内心历程。这历程影响句子结构的正确性，以及对其含义的理解。这个语言学的影响因子让认知心理学家共同在语言学习上找到新方法。也把原来附属于行为科学里的描述性语言研究引进认知科学中。所有上列构想和理论也都经过一段时间之后被重新发展和更新，直到奈瑟尔（1967）才把所有项目聚集在一块，定名为**认知心理学**。

1.5 认知心理学到认知科学

认知的基本是在考虑人处理信息的过程，即人类转移、减缩、叙述、储存、回收和利用"信息"的能力。"信息"在这方面的解释，就是"知识"。只要知识能做成数字呈现，而且能转化成计算机程序，则其运作过程就可在计算机中仿真出来，让机器有能力执行具有类似人类智慧般的专业课题。因此，受到计算机科学中人工智能的影响，并且配合认知心理学中对高层次心智活动和结构的研究，一门跨领域学科——**认知科学**，就成形出现了。认知科学的研究集中在了解心智呈现，分析思考，及以计算机模式来仿真人类思考（Gardner, 1985）。

下面简单的例子能较清楚地解释这一观念。比方说，人脑相似于一个大机器，它接受信息输入和信息输出。在人生存的周遭大环境中，环绕着无数有形及无形的信息。这些信息会经由一些媒介产生一些刺激。这些刺激（例如文字、影像、声音、味道及力量），经由感觉器官输入人脑。输入后，信息会经过一些知觉的运作放在短程记忆中，而后再经

图 1-12　司马贺（1916~2001）
Figure 1-12　Herbert A. Simon (1916~2001)

provides output. In our living environment, we observe large amounts of visible and invisible information. This information via certain media can generate certain stimuli (i. e., text, image, audio, taste, and forces) that can be put into the human mind through human senses. After this input, the senses transfer the information into short-term memory, convert the information to knowledge, and store the knowledge in long-term memory. Afterwards, the knowledge can be retrieved from our long-term memory, put into the short-term memory appropriately, and convert into information ready for output. The purposes of information output are to provide reactions for responding to given stimuli.

Realistically, when a stranger introduces him/herself by name, two pieces of information are provided. One is the image of the stranger's facial appearance; the second is the name label. These two pieces of external information will be encoded through visual and auditory sensory receptors and input to the short-term memory. "Encoded" means to convert the information into useful coding symbols. Short-term memory is a temporary information storage area. In where, mental process would convert the sound of the name to a symbol code and the facial appearance to an image code; and associate the two together into one knowledge pattern, and conditionally store it in long-term memory. At this point, knowledge is learned. It also will be conditionally, according to the new stimuli, determined to search for the related information from the long-term memory. If the information is found, then the name label, facial image, or related information would be sequentially placed back in the short-term memory for use. At this time, a reaction to the stimuli could occur.

The entire processes could be separated into three groups of questions: (1) the **cognitive mechanism** questions of how information is input, and knowledge is output; (2) how the knowledge is stored and the representation for storage related **mental structure** questions; and (3) the **mental processing** questions of how knowledge can be found and utilized. Combining these activities and processes together, a fundamental cognitive system is established. The analytical research of this cognitive system has to develop models capable on predicting certain phenomenon. These models should be built from setting up implicit hypotheses and by conducting repeatable and empirical experiments for verification. Yet, different individuals might have different cognitive performances. Because of this, more cognitive psychologists are needed for executing additional psychological experiments to find out acceptably firm models through statistical methods. More scholars from computer science and artificial intelligence are also needed to simulate the built models for applications. Thus, cognitive science is an interdisciplinary field with applications useful to various disciplines.

Note:

1. See Web page on Socrates, URL: http://en.wikipedia.org/wiki/Socrates
2. See Web page on Plato, URL: http://en.wikipedia.org/wiki/Plato
3. See Web page on Aristotle, URL: http://en.wikipedia.org/wiki/Aristotle
4. See Web page on St. Augustine, URL: http://en.wikipedia.org/wiki/Augustine_of_Hippo
5. See Web page on Nicolaus Copernicus, URL: http://en.wikipedia.org/wiki/Copernicus
6. See Web page on Galileo Galilei, URL: http://en.wikipedia.org/wiki/Galileo_Galilei
7. See Web page on Johannes Kepler, URL:

过一些运作，把信息变成知识，储存在长期记忆中。再往后，知识可随时在记忆中被寻找，在适当时候被放到短程记忆中转成信息准备输出，以便对环境中遇到的刺激作适当的反应。

举个实例说明：当一个陌生人介绍他／她名字时，有两个信息存在。一是陌生人的脸孔长相，另一是名字符号。这来自外界的两个信息，借着视觉和听觉两个知觉感官进行编码放到短程记忆中。"编码"之意是把外界信息转化成有用的心理格式符号以便运作。短程记忆是个信息暂存区。知觉会把脸孔长相化成影像，把名字的音符化成符号，并合并成一个知识组块，依照情况决定要不要存到长期记忆中。这时，知识已学到。知觉也可以对新来的知觉刺激，依照情况决定要不要在长期记忆中去寻找所要的数据。如果数据找到，这名字符号、脸孔影像或相关知识数据会依序传回短程记忆中以便运用，这时对刺激的反应就可能产生了。

这整个系统活动可分离成三大问题揽括整个历程：（1）信息如何被输入、知识如何被输出的**认知机制**问题；（2）知识是如何被储存和以何种形式储存的**心智结构**问题；（3）知识如何被寻找、使用的**心智过程**问题。把这些涉及的活动历程综合起来，即构成一个宏观的基本人类认知系统。分析整个认知系统的研究，需要建立起一些可以解释，以及可以预测现象的模型。这些模型要透过严密的假设和一系列可重复验证的实验之后而建立。然而，认知系统的表现会因人而异。因此，这学科也需要认知心理学家做许多心理实验，以统计方法找出一些普遍的解说模型。也需要人工智能学者和计算机科学学者共同把这些找出的现象及模型在计算机中仿真运用。所以这学科的研究是跨领域的，运用也是跨领域的。

附记：

1. 苏格拉底的资料网址：http://en.wikipedia.org/wiki/Socrates
2. 柏拉图的资料网址：http://en.wikipedia.org/wiki/Plato
3. 亚里士多德的资料网址：http://en.wikipedia.org/wiki/Aristotle
4. 圣·奥古斯丁的资料网址：http://en.wikipedia.org/wiki/Augustine_of_Hippo
5. 哥白尼的资料网址：http://en.wikipedia.org/wiki/Copernicus
6. 伽利略的资料网址：http://en.wikipedia.org/wiki/Galileo_Galilei
7. 开普勒的资料网址：http://en.wikipedia.org/wiki/Johannes_Kepler
8. 牛顿的资料网址：http://en.wikipedia.org/wiki/Isaac_Newton
9. 康德的资料网址：http://en.wikipedia.org/wiki/Immanuel_Kant
10. 罗素的资料网址：http://en.wikipedia.org/wiki/Bertrand_Russell。版权为罗素档案室、麦克马斯特大学图书馆所拥有。
11. 笛卡儿的资料网址：http://en.wikipedia.org/wiki/Rene_Descartes
12. 洛克的资料网址：http://en.wikipedia.org/wiki/John_Locke
13. 费希纳的资料网址：http://en.wikipedia.org/wiki/Gustav_Fechner
14. 赫姆霍兹的资料网址：http://en.wikipedia.org/wiki/Hermann_von_Helmholtz
15. 冯德的资料网址：http://web.lemoyne.edu/~hevern/nr-theorists/wundt_wilhelm.html

    http://psychclassics.yorku.ca/Wundt/Physio/wozniak.htm

    http://www.psy.pdx.edu/PsiCafe/KeyTheorists/Wundt.htm

    http://educ.southern.edu/tour/who/pioneers/wundt.

http://en.wikipedia.org/wiki/Johannes_Kepler
8. See Web page on Issac Newton, URL: http://en.wikipedia.org/wiki/Isaac_Newton
9. See Web page on Immanuel Kant, URL: http://en.wikipedia.org/wiki/Immanuel_Kant
10. See Web page on Bertrand Russell, URL: http://en.wikipedia.org/wiki/Bertrand_Russell. Photo courtesy is from Bertrand Russell Archives, McMaster University Library.
11. See Web page on Rene Descartes, URL: http://en.wikipedia.org/wiki/Rene_Descartes
12. See Web page on John Locke, URL: http://en.wikipedia.org/wiki/John_Locke
13. See Web page on Gustav Fechner, URL: http://en.wikipedia.org/wiki/Gustav_Fechner
14. See Web page on Hermann von Helmholtz, URL: http://en.wikipedia.org/wiki/Hermann von Helmholtz.
15. URL of Web pages on Wilhelm Wundt:
http://web.lemoyne.edu/~hevern/nr-theorists/wundt_wilhelm.html
http://psychclassics.yorku.ca/Wundt/Physio/wozniak.htm
http://www.psy.pdx.edu/PsiCafe/KeyTheorists/Wundt.htm
http://educ.southern.edu/tour/who/pioneers/wundt.html
http://en.wikipedia.org/wiki/Wilhelm_Wundt
http://plato.stanford.edu/entries/wilhelm-wundt
http://www.colorsystem.com/projekte/engl/23wune.html
16. See Web page on William James, URL: http://www.ship.edu/~cgboeree/wundtjames.html
http://en.wikipedia.org/wiki/William_James
17. See Web page on John Watson, URL: http://facweb.furman.edu/dept/psychology/watson1.htm
http://en.wikipedia.org/wiki/John_B._Watson http://www.pbs.org/wgbh/aso/databank/entries/bhwats.html
Photo courtesy is from the Archives of the History of American Psychology, University of Akron.
18. See Web page on Max Wertheimer, URL: http://en.wikipedia.org/wiki/Max_Wertheimer. Photo courtesy is from the Archives of the History of American Psychology, University of Akron.
19. See Web page on Wolfgang Kohler, URL: http://www.kirjasto.sci.fi/kohler.htm. Photo courtesy is from the Archives of the History of American Psychology, University of Akron.
20. See Web page on Kurt Koffka, URL: http://en.wikipedia.org/wiki/Kurt_Koffka. Photo courtesy is from the Archives of the History of American Psychology, University of Akron.
21. See Web page on Sigmund Freud, URL: http://en.wikipedia.org/wiki/Sigmund_Freud http://en.wikipedia.org/wiki/Freud
22. See Web page on Jean Piaget, URL: http://en.wikipedia.org/wiki/Jean_Piaget
http://www.learningandteaching.info/learning/piaget.htm
http://www.ship.edu/~cgboeree/piaget.html
Photo courtesy is from the Archives of the History of American Psychology, University of Akron.
23. See Web page on Ulric Neisser, URL: http://en.wikipedia.org/wiki/Ulric_Neisser
24. Hurbert Simon (1916~2001) had his Chinese name. See Chan, C. S. (2003) Thoughts of Herbert A. Simon – on artificial intelligence in Design. In Chiu, ML (Ed.), *CAAD Talks 2: Dimensions of Design Computation*, Garden City Press: Taipei, pp. 26~33. Copyright by Carnegie-Mellon University.
25. Resources of some photos shown in this

html

http://en.wikipedia.org/wiki/Wilhelm_Wundt

http://plato.stanford.edu/entries/wilhelm-wundt

http://www.colorsystem.com/projekte/engl/23wune.htm

16. 詹姆斯的资料网址：http://www.ship.edu/~cgboeree/wundtjames.html

http://en.wikipedia.org/wiki/William_James

17. 华生的资料网址：http://facweb.furman.edu/dept/psychology/watson1.htm

http://en.wikipedia.org/wiki/John_B._Watson

http://www.pbs.org/wgbh/aso/databank/entries/bhwats.html

图片版权为艾克隆大学、美国心理学历史档案室所拥有。

18. 韦特墨的资料网址：http://en.wikipedia.org/wiki/Max_Wertheimer。图片版权为艾克隆大学、美国心理学历史档案室所拥有。

19. 苛勒的资料网址：http://www.kirjasto.sci.fi/kohler.htm。图片版权为艾克隆大学、美国心理学历史档案室所拥有。

20. 考夫卡的资料网址：http://en.wikipedia.org/wiki/Kurt_Koffka。图片版权为艾克隆大学、美国心理学历史档案室所拥有。

21. 弗洛伊德的资料网址：http://en.wikipedia.org/wiki/Sigmund_Freud

22. 皮亚杰的资料网址：http://en.wikipedia.org/wiki/Jean_Piaget

http://www.learningandteaching.info/learning/piaget.htm

http://www.ship.edu/~cgboeree/piaget.html

图片版权为艾克隆大学、美国心理学历史档案室所拥有。

23. 奈瑟尔的资料网址：http://en.wikipedia.org/wiki/Ulric_Neisser

24. 司马贺的资料网址：http://en.wikipedia.org/wiki/Herbert_Simon。司马贺（1916~2001）是他的中文名。请见陈超萃. 人工智能与建筑设计：解析司马贺的思想片断之一——设计运算向度. 台北：田园城市文化事业出版社，2003：26~33. 版权为卡内基梅隆大学所拥有。

25. 本章图片取材自公众共享资源领域的维基词典，作视觉参考之用。维基词典的网页地址：http://en.wikipedia.org/wiki/Main_Page。中文网页地址：http://zh.wiktionary.org/wiki/Wiktionary:%E9%A6%96%E9%A1%B5。公平使用影像的合理性可见解说于网页 http://en.wikipedia.org/wiki/Wikipedia:Images

chapter are from the public domain of "Wikipedia, The Free Encyclopedia", for visual reference. Permission is granted to copy, distribute and/or modify the document under the terms of the GNU Free Documentation License. The URL of Wikipedia is:
http://en.wikipedia.org/wiki/Main_Page. The URL of its Chinese version is:
http://zh.wiktionary.org/wiki/Wiktionary:%E9%A6%96%E9%A1%B5. *FAIR USE of IMAGENAME.jpg: see image description page at: http://en.wikipedia.org/wiki/Wikipedia: Images for rationale.*

第2章 设计与认知的关联

设计可归类于所有人类有创造性地为满足一些需要而制造出一些物品，或为适应某些目标而作出一个结构体的努力。在专业的本质上通常要求考虑美、机能使用、社会表征和市场需求等因素。因此，所有人为设计的基本元素就是：设计都必定有设计意图（即原动力）在驱动，尔后经过一系列的动作执行，产生一些结果（产品）。如果把设计从构思直到成形结果的整个过程的角度来看，设计可分成自然设计和深思熟虑设计两类。两类都有一些实质上或观念上的结构体存在（Perkins，1986）。

自然设计（或直觉设计）是由某一盲目的自然程序发展出来的结构，而非经由某一位或某一组制造人，经过蓄意操作经营而形成的结果。例如，语言、风俗和时尚都是文化的产品。各自都有其特定的规律结构去主导成形过程，并且满足某一特定文化的特定需要。不过，文化的产品都是逐渐在日常生活中改变，经过约定俗成而来，并被人们适应，也被公众接受的。因此，文化改革演进如就扮演一种非正式性、集体性，而且是制造性的演进过程本身而言（Alexander，1964；Rapoport，1969），文化可看成是被社会演化打炼出来，被调整适应力驱动出来的一种自然设计产品（图2-1）。因此自然设计的驱动力是调整并适应，产品则是语言、风俗和时尚等。

深思熟虑的设计（或意图设计）是一位或一群制造者，经过计算之后为适应某种目的而做出的结构。通常在结构体完成之前，某位设计者会考虑构件，寻找构件所要满足的目的。有时在某

图2-1 设计种类及其创造力量
Figure 2-1 Diagram of design category and creative forces

Chapter 2 Correlations between design and cognition

Design can be attributed to all human creative endeavors of shaping objects to meet purposes or constructing a structure adapted to objectives; which require professional consideration on aesthetic beauty, functional uses, social symbols, and market demands and supplies. All man-made design, thus, has the fundamental essence that it is driven by certain intentions (which are the driving forces) and achieved by a series of actions to generate results (which are products). In this regard, design can be categorized in two groups, natural and deliberate; both have some physical or conceptual structures associated with (Perkins, 1986).

Natural designs or instinctive designs are structures developed by a blind natural process and without the calculated shaping of any group or individual maker. For instances, language, custom, and fashion are cultural products which have rule structures governing the formation process for suiting better the needs of a culture. Yet, these cultural products are gradually changing in everyday life, adapted by people, and accepted by the public. The cultural evolution as an informal, collective, and generational process (Alexander, 1964; Rapoport, 1969) is a natural design product forged by social evolution and driven by the force of human adaptation (Figure 2-1). The driving force in natural designs is adaptation, and the products are language, custom, and fashion, etc.

Deliberate or intentional designs are structures made suitable to a purpose by the calculating act of a maker, or group of makers. A designer conceives the structure and its needed purposes before the structure is completed. Individuals working as a group may also design a structure to match some purpose gradually over time. Artificial products made to suit specific intentions, computer programs developed to run particular tasks are deliberate designs. In Fine Arts, the selection and organization of materials, utilities, visual elements, and esthetic principles to achieve a desired effect is deliberate design. In architecture, the social process that responds to the needs of users who work and live in the building, or the process that concerns aesthetic manipulations and expressions are deliberate design. Put the meaning of design in a larger scope and wider context, all occupations engaged in converting actual to preferred situation are concerned with design (Simon, 1969), which are driven specifically

个过程中的某种情况下，也会有一群设计师合力共同完成一结构，逐渐依时体现既定的目的。比方说，为某些特定功能而做的人工产品，成为执行某种特殊课业而设的计算机程序等，都是深思熟虑的设计。在美术设计中，选取材料、工具、视觉因素及美学原则，并把全部组织起来以达到预期的效果，即是深思熟虑的设计。在建筑中，对居住在建筑物里的居民和工作在建筑物里的使用者之心理或实际要求的考虑而产生的社会对应之过程；或关切美、运作美并表达美的所有过程也都是深思熟虑的设计。把这些观念放在大格局中来看，则所有转化事实情况到理想情况的行业都涉及设计（Simon，1969）；而领军的是各领域的专业知识。因此，深思熟虑的设计之驱动力是设计意图，而产品则是人造物。

理论上，如果把设计当作是由某种特殊驱动力量应运生成的智慧冒险结晶，则有三个根本层次值得探讨：一是设计规则；二是设计方法论；三是设计思考过程（图 2-1）。这三个层次都与设计的本质密切相关，而且是生成有质量产品的根源：

（1）设计规则可视为设计时使用的机械原则，是被公众认同、可依循重复使用、开放性之设计准则。这准则可由产品中体会出结果，也可被归为设计中原则性的方法类。例如，空间比例、物体尺度大小、材料颜色、空间流通或建筑法规等都是基本的设计原则。

（2）设计方法论是以一些理论框架为主导的系统化程序学说，使用在设计中。例如，史迹维护及能源节约是主要的设计观念，也需要特殊考虑其设计方法。

（3）至于设计思考过程，则是一个设计由作草图到完工的整个思考历程。这历程可以说是设计师处理一设计案的个人内在设计认知旅程。

实务上，设计师会认清设计课题，察觉要针对的设计问题，使用一些规则确定要寻找的设计数据，利用设计方法生成、评估，并选定结果。其中，设计规则和方法来自学习的设计知识，学习本身就是人类认知的一部分。至于设计过程，在另一方面，指的是执行认知的操作过程。整体而言，这三个设计本质的层次即是在设计中所实现的认知之成果或现象，总体定名为"**设计认知**"。本章将把设计规则、设计方法论和设计思考过程三个主要层次中与认知有关的研究结果作一叙述性介绍。

2.1 设计原则及方法

设计方法与创造好产品的设计原则及步骤有关。不同产品，使用的设计手法各异。但同种产品，如在制造过程中使用不同程序手法，结果自然也会不同。设计者所关切的及有兴趣的，是寻找好的步骤技巧以及培养出判断好产品之能力。这些都与前章中所讨论的心理学中视知觉与知识获取有关。市场上设计业很多，包罗万千，不克详举。但热门且熟知的设计业被认为是建筑设计、都市设计、工艺设计、时装设计、图案设计、工业设计和机械设计等。这些专业中，工艺设计和图案设计的设计方法是最早受到心理学影响的专业。影响的学派是完形心理学派。

2.2 完形心理学与设计

完形心理学专注于了解从感官中得到的新知觉，是如何被组织成一个综合及整体感受的。完形心理学的创办人之一韦特墨所设立的**场学说**（Wertheimer，1923；Koffka，1935）及视觉认知，在视觉艺术设计及工艺设计的专业里（包括建筑、手工艺、工业设计、绘画、雕塑、戏剧及印刷）有些影响。场的观念认为，知觉世界里也有一种类似磁场及电场的视觉场存在。这场学说与阶级学说有些不同，阶级学说形容个体在特定阶级的归属地位，决定了该个体的行为，即阶级决定个体行为。但场学说，换个角度，明示物体的行为是决定于它们所归属的场之结

by certain domain specific knowledge. The driving force of deliberate design is design intention, and the products are the artifacts.

Theoretically, if design is seen as the intellectual venture and products as generated by specific driving forces (Figure 2-1); there are three fundamental layers——design principles, design methodologies, and the design thinking processes——worthy of study. These layers relate closely to the nature of design and are the basis for creating qualified design products.

(1) Design principles are mechanical rules used in designing and commonly recognized as design guidelines capable to be followed and utilized publicly. These guidelines could be perceived from products and categorized as types of principle methods. For instance, proportion of spaces, size of components, color of materials, circulation between spaces, or even building codes are fundamental design principles.

(2) Design methodologies are theories of systematic design procedures supported by certain conceptual frameworks applied in designs. For example, historical preservation and energy preservation are major design concepts that need special considerations and methodologies.

(3) Design thinking processes are the entire thinking experience starting from the conceptural design stage to the project completion stage. The processes are intrinsic design cognition experienced by the individual designer when handling a design project.

Practically, designers recognize design tasks, perceive design issues, and apply principles to search for design information; and utilize methodologies to generate, evaluate, and decide results. Design principles and methodologies are design knowledge developed from learning, which is a part of human cognition. The design process, on the other hand, means the operational procedures of executing cogni-

tion. Overall, these three layers of the design nature are, to some extent, either the results or the phenomena of performing cognition in design which should be termed "**design cognition**". This chapter introduces notions and concepts from cognitive perspectives, particularly, works related to design principles, methods, and thinking processes.

2.1 Design principles and methods

Design methods relate to the principles and procedures for creating good products. Different products have different generational methods. But for the same types of products, changing the produce process changes the results. Designers' concerns and interests focus most on quality techniques and judgments. These considerations are related to the previous Chapter's discussion of visual perception and knowledge acquisition in the field of psychology. There are a large number of design fields in the market; the familiar and popular ones are architectural design, urban design, craft design, fabric/fashion design, graphic design, industrial design, and mechanical design, to name a few. Craft design and graphic design were strongly influenced by Gestalt psychology.

2.2 Gestalt psychology and design

Gestalt psychology involves recognizing how sensations are organized into a unified perception, or whole. Professions of visual art design and craft design (including architecture, handicrafts, industrial design, painting, sculpture, theater; and typography) have been influenced by the **field theory** (Wertheimer, 1923; Koffka, 1935) and perceptual cognition developed by Max Wertheimer, one of the founders of Gestalt psychology. In field theory, a field is similar to a magnetic or electronic field, existing in our world

构，或者说是以视觉能量在场中的时空安排布局而定（Hartmann，1935）。场会是一个景或一幅画。人对物体的知觉是由它存在于场中的状态或条件而决定的（Koffka，1935）。因为场中元素的形成可能是整体也可能是散漫个体。完形心理学的认知观是描述在视觉场中如何对整体形成认知（Ellis，1999）。因此，场地论是代表完形心理学的观念之一。

场学说观念被**包豪斯设计学院**的教授认同（Scheidig，1967；Gay，1976）。包豪斯是沃尔特·格罗皮乌斯于1918年在德国魏玛成立的艺术及工艺学校。格罗皮乌斯在创校时就相信包豪斯的教育不应只局限在实际操作"机能"。它应有以心理学做科研调查而发展出的"美术哲学"存在。建筑师、画家及雕刻家应认清建筑物完美多重形里的整体性和个别性（Gay，1980）。包豪斯并没直接引用完形心理学，但包豪斯对视觉场中视觉元素与元素间的数字性和表现性的分析有过研究。瑞士艺术家保罗·克利于1921年加入包豪斯后，就集中研究感觉和知觉，并把完形心理学中的组合原则用于课程教材中（Teuber，1973）。柏林完形心理学校不但在包豪斯开了一些课做些教学合作，而且完形心理学对整体的强调，也促成包豪斯结合了**英国美术工艺运动**（1880~1910）的观念把建筑结构理论放在建筑设计课程里，把艺术和科学融合，并结合工艺生产和实践，引出一个新的设计方向。这方向奠定了现代建筑运动的起步。

完形心理学从精神物理组合的角度提出一个视觉上的**简单完好律**来解释在视觉场中的视觉认知因果。简单完好是意译，原文是德文的怀孕之意，但在完形心理学中的用法是指孕育出知觉意义。完形心理学认为人的视觉是不看不相关的光线颗粒，而看物体、对象和形态。比方看一页字，看到的不是字母而是字母组成的单词（Kohler，1930）。这观念明示在视觉里，真实会被组织起来，并根据规则性、对称性和简单性，把图形简化成最简单的形。例如图2-2会被看成是一系列的圆圈而不是许多复杂的图块。为了解释整体和整体中个体的视觉关系，完形心理学也提出几个**视觉组织规律**或原则（Wertheimer，1923；附记1），解说小构件是如何在视觉上组成大构件的。而且解说视者如何去看组件而得出整体的意义。完形心理学认为见树是见不得林。因为如只看到个体（树），个体（树）的特色并不完全代表整体（林）的真相。下列六个定律很妥切地解释了视觉场中的视觉认知现象。

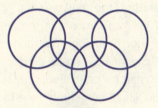

图 2-2 简单完好律
Figure 2-2 Law of pragnanz

（1）**接近律**：物体以等距离靠近的（邻近或贴近）会被看成是聚在一块的组体。图2-3中左边的圆形物体会被看成是垂直柱式，而右边的三角体会被看成是水平列条。图2-4证明接近律是如何把12个物体在视觉上组成三组，而不是12个单元个体。

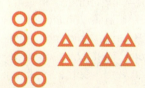

图 2-3 接近律例 A　　图 2-4 接近律例 B
Figure 2-3 Example A　　Figure 2-4 Example B
of proximity　　　　　of proximity

（2）**相似律**：在某些情况下，相近律是组成视觉形块的首要因素。但当物体邻近距离逐渐加大到某一个程度时，相似律就变成主要决定因素了。相

of visual sense. Field theory differs from class theory, which indicates that the membership in a certain class determines the behavior of an object, or the behavior of objects is determined by the class to which they belong. Field theory means that the behavior of an object is determined by the field structure or spatio-temporal configuration of the energy within the field (Hartmann, 1935). The field could be a scene or a picture, and human perception of the object would depend upon how the condition or situation of that the object exist in the field (Koffka, 1935). Elements in the field could be grouping or not grouping. The cognitive aspect of Gestalt psychology is to describe how the whole group can be perceived cognitively (Ellis, 1999). The visual perceptual field theory is one of the theories in Gestalt psychology.

The concept of field theory was accepted by the faculty at **Bauhaus** (Scheidig, 1967; Gay, 1976). Bauhaus, the School of Art and Craft, was established in Weimar in 1918 by Walter Adolf Gropius. Gropius believed that the educational principles in Bauhaus should not be limited to the practical operation on function, but should include esthetic philosophy based upon psychological investigations. Architects, painters, and sculptors should recognize and grasp the multiform shape-vielgliedrige (in German) Gestalt of the building in its totality and its parts (Gay, 1980). Bauhaus did not apply Gestalt psychology directly, but conducted research on mathematical and representational analysis of the field elements. The Swiss artist Paul Klee joined Bauhaus in 1921, including the study of sensation and perception, and applied the laws of grouping in his course materials (Teuber, 1973). The Berlin School of Gestalt Psychology had contributed to several courses with Bauhaus, and their influence led Bauhaus to incorporate concepts from the **Arts and Crafts Movement** in England (1880~1910), including structural concepts in their architectural design studios, integrating arts and science, and combining crafts and practice to create a new architectural design direction. This collaboration was the basis for the modern architecture movement.

Gestalt psychology proposed the **Law of Pragnanz**, applying the casual-effects of cognition in the visual field from the psychophysical organizational point of view. Pragnanz is German for "pregnant", but its definition in the context of the Gestalt theory is the sense of pregnant with perceptual meaning. Gestalt psychology believes that human beings do not see unrelated bits of light; instead people see objects, things, and patterns, i.e., a group of letters that form a word (Kohler, 1930). The notion is that reality is organized or reduced perceptually, based on the properties of regularity, symmetry, and simplicity, to the simplest form possible. For example, the image in Figure 2-2 is perceived as a series of circles rather than as a number of complicated shapes. Gestalt psychology developed several **perceptual organization laws** or principles to illustrate how smaller objects could be perceptually grouped to form larger ones, and how viewers could perceive meaning from the whole (Wertheimer, 1923; see note 1). To them, seeing trees could not comprehend the forest. Parts (trees) identified individually would have different characteristics to the whole (forest), besides, parts (individual trees) could not represent the characteristics of the whole (forest). The following six laws define visual cognition in the visual field.

(1) **Law of Proximity**: Objects close to each other with equal interval space between them (proximal or contiguity) tend to be perceived as a unified group. The left part objects

似律指出相似的物体倾向于被组成一块。例如图 2-5 中图形会被看成是一组垂直红圆,一组红三角,一组棕色圆,一组双心圆和四组垂直图案方块。相似律中会因相似性而形成一组的因子,包括相似的尺度大小、形状、色彩、质地和物体方向等。

图 2-5　相似律
Figure 2-5　Law of similarity

（3）**共同命运（单一终点）律**：如果相近律和相似律都已用到图形里,则单元物体有相似方向和节拍的物体会在视觉上被组成一体,产生运动感。例如图 2-6 的下部图形有往下垂的视觉效果。

图 2-6　共同命运（单一终点）律
Figure 2-6　Law of common fate

（4）**好的连续律**：视觉上,直线是连续的安定线,曲线是动线。而视觉会把曲线沿着最顺的曲率作视觉延伸。图 2-7 的下部线自然会被看成是左半部的顺延。这也表示视者有接受做成最好曲线的视觉效果倾向,而非把图案打破成多重碎块。人类视觉上有好的连续图形的倾向,也会将图 2-8 看成是两根曲线 AC 及 DE,而非四根。

（5）**结束律**：视觉上有倾向把不完整的刺激信息打造成完整的美好结束图形。视者会忽略断点残缺而自动地把边线补全化作完整的轮廓。例如图 2-9 中没有完整的三角和圆形,但心智会创出缺失的边和角,化不完整成完整的熟悉图形。图 2-10 中的三个字母 I、B、M 会被认出是字母,即使形状是由不同长短白蓝水平线条互相重叠在一起的。

图 2-9　结束律
Figure 2-9　Law of closure

图 2-10　IBM 商标（1972 年版权）
Figure 2-10　IBM logo（Courtesy of IBM, 1972 copyright）

（6）**区域与对称律**：区域原则是说如有两个图形重叠,特别是一个图形出现在一个均匀的场中时（Wertheimer, 1923）,则小的将被看成是正场,是被包含的形体或称图形,而大的是负场,是会要包含的形体或称地景（图 2-11）。要认出二者之一就要依注意力和视线固定点而定。这观念最早被鲁宾（Rubin, 1915; Hartmann, 1935）于 1914 年做的实验得到证明。对称的原则指出当我们在看物体时,我们会倾向于看它们是一个围绕中心的对称形体。在看对称形时,整个图形会被观察到,而非只是单独构件。从图 2-12 看到的是两个钻石形,而非一个小钻石加上两个残缺的方块。对称图案的原则是对

图 2-7　好的连续律例 A
Figure 2-7　Example A of continuity

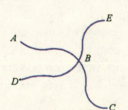

图 2-8　好的连续律例 B
Figure 2-8　Example B of continuity

in Figure 2-3 are grouped in vertical columns while the right part of objects appeared to be in horizontal rows. Figure 2-4 exemplifies how the principle of proximity has visually grouped the twelve items into three groups as opposed to twelve individual items.

(2) **Law of Similarity**: Proximity is the predominant factor in some cases in visually banding objects together, but a gradual increase of intervals will reach the point at which similarity is predominant. Similarity indicates the tendency of like parts to band together. For instance, people would see objects in Figure 2-5 as a group of vertical red circles, a group of red triangles, a group of brown circle, a group of inner circle, and four vertical groups of textured rectangles. Common factors that affect grouping by similarity are similar size, shape, color, texture, and orientation of objects.

(3) **Law of Common Fate** (Uniform Destiny): If the principles of similarity and proximity are employed, then objects moving in the same direction at the same rate will be grouped visually together and a movement occurs perceptually. For instance, the lower part of Figure 2-6 appears to be moving downward.

(4) **Law of Good Continuation**: Visually, a straight line is a continuous straight line, whereas a curve line is a moving line. Human vision will perceive curved lines by following the smoothest path. For instance, the bottom branch in Figure 2-7 is seen as continuing the first segment of the line. This allows people to see figures as flowing smoothly without breaking lines up into multiple parts. The preferences for continuous figures would have perceived Figure 2-8 as two curved lines of AC and DE instead of four.

(5) **Law of Closure**: Perceptually, there is the tendency to make complete, whole figures out of incomplete stimuli for a good end result. Viewers would ignore gaps and automatically complete the contour lines. In Figure 2-9, there are no triangles or circles, but the minds would fill in the missing information for creating familiar shapes and images. In Figure 2-10, the letters I, B, and M are perceived as characters, even though the shapes are combined with different length of white and blue horizontal lines that are hovering each other.

(6) **Law of Area and Symmetry**: The principle concept of area indicates that when two figures are overlapping, especially when one appears upon a homogeneous field (Wertheimer, 1923), then the smaller one is perceived as the included, positive field, or called figure; while the larger one is the including, negative field, or called ground (see the left two figures in Figure 2-11). The perception of either one depends upon the attention and visual fixation. The original concept had been proved by Edgar Rubin through experiments done in 1914 (Rubin, 1915; Hartmann, 1935). The principle of symmetry means that we tend to perceive objects as symmetrical shapes that form around the center. When we see a symmetrical shape, the whole figure is perceived rather than the individual parts making up the whole. In Figure 2-12, we see two diamonds rather than one diamond plus two broken squares. The principle of a symmetrical figure also means that the composition will be seen as a closed figure and the symmetrical contours will define a figure isolating from the ground (Figure 2-13). In this case, stronger contrast of color will show clearer visual effects (comparing the black-and-white verses red-and-white compositions in Figure 2-13). Black-and-white has stronger effects than red-and-white.

Based upon these organizational principles, in order to analyze the fixation and recognition of figures, a **Law of Figure/Ground** (Rubin called

图 2-11　区域律　　　　　图 2-12　对称律
Figure 2-11　Law of area　　Figure 2-12　Law of symmetry

称图形将会被看成是一个完整的形体，而且其对称轮廓线会将这图形从地景中分离出来（图 2-13）。在这情况中，色彩对比越强烈，则视觉效果越清楚。比较图 2-13 中黑白和红白的图形，可知黑白比红白效果更强烈。

图 2-13　对称图形　　　　图 2-14　图形与地景的分离律
Figure 2-13　Symmetrical figures　　Figure 2-14　Principle of figure/ground

根据这些规律，为分析视觉如何固定并辨认图形，一个**图形与地景分离**的原则被提出。这**图形与地景的分离律**（鲁宾称之为惯性姿态的结果）是说人的视觉场景通常会被区分成图案和地景两个基本部分。在看一个景时，有些物体会扮演较强的（图形）角色，其他的物体会退到地景（背景）去。例如著名的麦金塔计算机公司的徽章，就可看成是一幅正常的笑脸和一幅侧脸。应当注意的是如转换视觉焦点，则景色也跟着换。因为视者无法同时探出并领会图形和地景两种物体。如果焦点转移，地景就会变成图形，图形也变成了地景。比方视者把视觉焦点由图 2-9 中红色圆和三角（图形）转移到中间白色区（地景），则一个白三角即会在红色物体中

浮出。图 2-14 的正方体线框图，左下或者右上方块会被看成是方块顶部的视觉，取决于何者先被认出是图形。图 2-15 图形可是烛台也可是两张脸孔。

图 2-15　烛台还是脸孔
Figure 2-15　Candle holder or faces

图形与地景分离律比较适用于二维的图形和均匀的地景。如果图形是三维图案，则三维提供的景深会把视觉注意力减低（图 2-15 右图）。如果地景是不均匀的背景（图 2-16 左图），或色调强度对比不大（图 2-16 中图），或色素对比不强的颜色（图 2-16 右图），则分离效果会减弱。图 2-17 中三图，地景的物体是粗糙面各带红、绿、蓝色。粗糙面会

图 2-16　不同背景图纹的图形与地景分离律
Figure 2-16　Principle of figure/ground with different background texture

图 2-17　不同背景色彩的图形与地景分离律
Figure 2-17　Principle of figure/ground with different background color

it the products of a habitual attitude) was observed. The concept indicates that our visual field is usually divided into two basic parts of figure and ground. In perceiving a visual field, some objects take a prominent role (the figures) while others recede into the background (the ground). For example, the famous Macintosh computer company logo can be viewed as a normal happy face and a face in profile. Note that the change of view focus will also change the perception of the view. Because: viewers cannot comprehend both the figure and ground at the same time, the ground will become the figure when the focus is shifted. If viewers change the focus from the figures of black triangles and circles (the figures) to the ground in the center part (the ground) of Figure 2-9, a white triangle sitting on top of the black objects emerges. The lower left square or the upper right one of the wire frame of the cube in the left of Figure 2-14 could either be seen as the top face depending upon which one is viewed as the figure component. The image on Figure 2-15 could be viewed as either a candle holder or two faces.

The figure/ground law is most suitable for two dimensional figures and homogenous ground. If the figure part is three dimensional objects (the right figure in Figure 2-15), its visual depth clue would affect the power of recognition. If the ground part is not homogenous (the left image in Figure 2-16), or the contrast (saturation) between the figure and ground is weak (the middle image in Figure 2-16), or the values of color hue between the ground and the figure are close (the right image In Figure 2-16), results would be weakened. In Figure 2-17, the ground objects of these three images are bumpy with red, green, and blue colors associated, as such, the depth clue provided from both the ground part and the figure part do affect the visual attention and, thus, the figure/ground law is not obvious; regardless the symmetrical lighting shown on the figure part (compare the candle holder images in Figure 2-16 and 2-17). The two images in Figure 2-18 are three dimensional stereoscopic images, which should be seen through a pair of red (left eye) and blue (right eye) 3D viewing glasses to bring out the 3D effect. Image on the left of Figure 2-18 has black ground, in which the figure and ground effect is easy to recognize. Yet, in the image on the right with rough 3D background and 3D stereoscopic effect, the law of figure/ground does not apply.

In sum, the core notion of all studies done by Gestalt psychologists centered on the core phenomenon that " humans could mentally compensate missing information in the two dimensional scene to form a complete and good shape. The goodness results from the perceptual tendency to get the simplest figures possible. In other words, under the principle of simplicity, perceptual organizational processes form the best and simplest interpretations of sensory data. " Yet, the change of focus would switch the ground to figure which also would create surprising visual clues on comprehending the scene. In Figure 2-9, a white triangle emerges in front of the lines and incomplete circles, because the incomplete figures suggest that something is covering them (Rock, 1983). Effects of these perceptual organization laws and principles, especially the law of figure/ground in particular, have often been used by scholars to analyze art (Koffka, 1942; Teuber, 1974) to get understanding of architecture, music, painting, poetry, sculpture, cinema, and theater (Arnheim, 1966; 1974); and by graphic designers to design company logos. These examples illustrate the influences of cognitive psychology on design principles.

产生景深，因此地景的景深再加上图形烛台三维的景深，两个景深配合起来就削弱了视觉注意力。使得图形与地景分离律就变得更不明显，遑论图形本身有对称的光彩图画效果（比较图 2-16 和图 2-17 的烛台效果）。图 2-18 是三维影像。戴上红（左眼）蓝（右眼）双色的三维立体目镜可看出立体的烛台。在图 2-18 例子中，左图的地景是纯黑，视觉干扰不强，图形和地景还可辨认。但右图有绿色粗糙面背景，加上三维效果，图形和地景分离律就不明显了。

图 2-18　三维立体影像有和匀（左）及粗糙面背景（右）
Figure 2-18　Three dimensional stereoscopic images with homogeneous (L) and rough ground (R)

括言之，完形心理学所做的知觉研究和观察，围绕着一核心观念："即人会把二维图案中缺失的数据，以心智补全，完成一个图案。也为了完好，心智会倾向于把图形转化到最简洁的程度。即视知觉的转化过程会把视觉捕捉到的信息做到最简洁、也最理想的诠释。"然而，视点的转换会把图形和地景对调，而产生意想不到的意外视觉结果。例如图 2-9 中的白三角会浮现在断线三角和不完整的圆形上，是因为残缺的图形建议一个遮盖物体的存在（Rock，1983）。这些视觉组织规律所产生的独特效果，特别是图形和背景分离律这一项，已被学者应用来分析艺术（Koffka，1942；Teuber，1974），了解建筑、音乐、绘画、诗、雕塑、电影以及戏剧（Arnheim，1966；1974）；也被图案设计师运用在公司的标志设计中。这些例子说明了认知心理学对设计原则与方法的影响。

2.3　设计方法论

设计可被严格地定义为："把抽象本质、形和体，全方位赋予一些适当目的之有意图的过程。"每个设计的过程都包含许多心智活动期，有些活动对产品的制成有切身关系（Broadbent，1969）。也因此，设计的过程在 1950 年代，吸引了不少方法论的研究。当时，一些研究致力于发展一些系统化设计方法去有条理地经营整个设计程序，以及发展一些系统化的设计技巧以便适用于这些过程也都逐渐浮现出来。这些研究是取材于当时发展出的**解题模式**及**运筹学**（又称作业研究）两个领域之观念。解题模式这领域是一个以科学化的方法分析人类如何解决问题的研究领域（详情请见第 5 章）。运筹学则是利用数学模型和计算法有效地解决生活中的复杂问题。

设计方法论最早的理论由琼斯和亚历山大发展而出。两人于 1962 年在伦敦举办的"设计方法研讨会"发表了两篇杰出论文（Jones，1963；Alexander，1963）。这两篇论文都收集在《设计方法论之进展》一书中（Cross，1984）。琼斯认为自 1950 年以后，在计算机、自动控制及系统化的影响下，设计应该比传统的凭直觉和经验来思考更应要倾向于靠逻辑和系统化的思考。他指出，一个设计过程有"分析—综合—评估"的循环期（图 2-19）。一个设计方案是在着手的问题被分析过，需要的情报数据被收集过，并综合过之后而形成的。而后所产生的解答方案也要被评估，以测试其可行性。如果生成的方案不达满意的标准，这过程会重新开始。这种设计程序会循环地重复出现，一直到设计者达到一个可接受的最终方案为止。他的论文就详细地列出"分析期"过程中收集设计数据的方法及适合分析数据单元的因素和分析规格，"综合期"中为创出解决方案所需要的综合信息法，以及"评估期"中为评审解决方案所准备的标准条件等。整个过程都有特

2.3 Design methodology

Strictly speaking, the operational definition of design is "*the intentional development of fitting abstractions, shapes, and forms in all dimensions into purposes.*" This process consists of a number of phases of mental activities, some of which are critical to the generation of products (Broadbent, 1969; pp. 15 – 21). Design processes attracted a number of studies on design methodology in the 1950s. At that time, the developments of systematic design procedures for the overall management of the design process, and systematic design techniques to be used within the process were emerging. These studies emerged from the fields of **problem solving** and **operational research**. Problem solving is a field that scientifically analyzes how human beings solve problems. Details can be found on Chapter 5. Operational research uses mathematical models and algorithms to effectively solve everyday complex problems.

The earliest theories of design methodology were developed by J. Christopher Jones and Christopher Alexander. Both presented seminal papers at the Conference on Design Methods, London, 1962 (Jones, 1963; Alexander, 1963). These papers were compiled in "Developments of Design Methodology" (Cross, 1984). Jones proposed that under the influences from computers, automatic controls and systems, design should be treated more logically and systematically other than traditional methods based on intuition and experience. He pointed out (1963; 1970) that a design process has the cycle of analysis–synthesis–evaluation (Figure 2-19). A design solution is generated after the addressed problem is analyzed and information is collected and synthesized, then the generated solution will be evaluated to test its feasibility. If the generated solution is not satisfactory, the process starts over again. Such design procedures repeat cyclically until the designer reaches an acceptable final solution. His paper provided details on data collection methods, factors and specification of data items appropriate for analyses in the analysis period; methods for synthesizing information to create solutions in the synthesize period, and criteria used for evaluating solutions in the evaluation period. The entire processes have specific procedures and techniques applied. At the data analysis stage, he applied graphic theory and matrix to clearly identify the interactions between data items relating to the design problem for getting correct solutions (Jones, 1963).

Alexander explained that a design has the process of analysis–and–synthesis (Alexander, 1964). He also developed a concept of **pattern language** (Alexander, Ishikawa & Silverstein, 1968) which has a hierarchy of parts linked together by patterns. A pattern is essentially a morphological law, a relationship among parts within a particular context. Each pattern solves generic recurring problems associated with the parts. Patterns are not isolated but linked by order that applies to a design. The structure of the language is composed by links from larger patterns to small ones, which create the network of the language. Each pattern has a title, and collectively the title forms a language for a design (see note 2).

The entire operational concept of pattern language is a complex task. In his first example (Alexander, 1963) which tended to work on decomposition techniques, he applied the computer method of graph theory for implementation. The sequence starts from listing a number of requirements from the design problem context which could be the natural, psychological, economical, or social needs. Then, depending on whether a pair of requirements is connected

定的程序及步骤。尤其在分析数据的阶段中，他用图形理论和矩阵法明确定出与设计问题相关数据单元间的关系，以便得出明确的解答方案（Jones, 1963）。

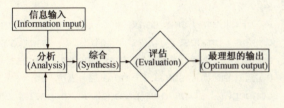

图 2-19 循环的设计过程
Figure 2-19 Cycles of design processes

亚历山大则解释设计是分析和综合的过程（Alexander, 1964）。他也发展出**模式语言**的概念（Alexander, Ishikawa & Silverstein, 1968）。模式语言由一套个体单元以某种形态有体系地连成一个整体。模式基本上是形态学上的一种规律，也是个体单元之间在环境中的特定脉络关系。每个模式也可能是解决一般会重复出现的问题答案。但模式本身并非独立无关，相反，模式是由与设计相关的设计单元以秩序相连接。整个结构即由大模式到小模式串连而成，这整个组构即形成这语言的网络。因为每个模式各自有其名衔，综合所有名衔就形成一个设计的语言（见附记 2）。

整个模式语言的观念操作是个庞大复杂的课题。亚历山大的第一个例子（Alexander, 1963），倾向于使用分解式的技巧，是用计算机图形理论法在计算机中实地操作。其程序开始是列出一系列与设计问题纵横相关的需求项目，诸如与自然界，与心理的、经济的，或与社交上牵涉到的设计要求。然后把要求项目配对，比较每一对要求项目间的相互关系是独立还是相依来决定他们之间是否相连。从计算机的图形理论方面解释，每个要求项目可当作是一个点，一对要求项目间如存在相依性则联机就存在。主要的课题是要把所有的要求项目和联机以区划法

减缩到独立的子集合。这子集合就可以数学定出一个线性图形代表这机能环境的结构。得出的图形结果也就定位了这环境的所有次要系统。剩下的工作就要做设计把要求单元与次要系统配套。在他的印度村庄例子中，他列出 141 个基本要求，以数学组成 12 个次要子系统，再更进一步组合并成 4 个主要系统，最后以图形设计安排这 12 个次要子系统单元、4 个主要系统单元和整个村落的结构。这整个村落的最终组织结构图形也就代表了许多可能的都市计划解决方案之一。

除了应用图形理论之外，设计方法论的研究方向也曾经由经营管理科学、运筹学和问题解决理论这三方面来着手。例如阿澈在 1970 年就运用这些不同学科的技巧，提出了一个系统化模式和逻辑运作法，把整个设计过程定出一个全盘化完整架构。在他的模式里，他把设计行为划分成八组特殊的设计动作群。这些设计动作先后各自相连。一个设计动作的发生或其本身状态的启动变化是导因于其前一个动作的发生或其状态变化而引起。因此，线性规划和**模控学**（或译神经机械学，见附记 3）的方法就被用来将设计过程数学公式化。这整个数学程序就是他所提出的系统模式的主要结构部分。在每一个划分出的设计期里，他也提出对应的逻辑运作，包括不同的设计工具和系统化方法等。这些逻辑是用来操作推动设计过程以便往前移动。这研究隐征性地解释了设计过程有三个成分：（1）设计动作是有序地依时往前进行；（2）设计过程会从解决一个主要设计问题过程中分叉到解决一些包含的子问题；（3）重复一些例行程序去循环性地进行问题解决（Archer, 1970）。

顺着对设计过程相似的论证，泽瑟（1981）也隐喻设计过程不是线性进行而是螺旋性进行的（图 2-20）。整个过程中有想像、表达和试验的三个单元重复运作期，而产生观念变换和产品发展的结果。想像是为某一特定问题而建构出一个心智影像的内

or independent, the relationships between pairs are used for creating links. Explained from the computer graphic theory, the requirements are treated as points, and the dependences between requirements are links. The major task is to decompose the number of requirements and number of links by partition them into independent subsets which mathematically define a linear graph representing the structure of the functional environment. This result, of course, creates a set of subsystems of the environment, what left are the components to be designed for fitting in the subsystems. In the Indian village example, he listed 141 basic requirements, which are mathematically grouped into twelve minor subsystems, and further combined into four major subsystems. Finally, he graphically demonstrated the components and the entire village, which was one of many possible design solutions.

Other than applying graphic theory, studies on design methodology also had been approached from the fields of management science, operations research, and problem solving theory. For instance, based on the techniques of these fields, Archer had proposed a systematic model and logic operations to frame the overall structure of the design process. In his model, he classified the act of designing into eight groups of unique actions. These actions are all related. One action occurs or changes its state by having the previous action or state changed. Thus, linear programming and **cybernetics** (see note 3) methods were used to formulate the design progress, which were the major parts of the structure of his proposed systematic model. Within each of the divided design phases, he also provided various design tools plus systematic methods to be used as logic operations for moving the design process ahead. The study had metaphorically explained that the design process had three major components of: (1) moving design actions ahead through time; (2) branching the design process from major problem into sub-problems; (3) reiterating routines to cyclically process problem solving (Archer, 1970).

Following the same line of reasoning the design process, Zeisel (1981) had also metaphorically indicated that design process is not lineal but spiral (Figure 2-20) on which conceptual shifts and product development are results of repeated iterative movement through the activities of imaging, presenting, and testing. Imaging means the internal process of constructing a mental image to fit specific pieces of a problem. Mental images are the products of applying subjective knowledge to organize ideas needed for a particular design problem. Presenting is utilizing all possible ways to externalize and communicate the designer's mental image presenting the design solution. Testing is to judge the feasibility of the generated image (solution) or to refine the generated solution to improve the presentation. These three elements of imaging, presenting, and testing are the mental activities constituting a design process.

According to Zeisel, a designer usually is backtracking to look backward for determining how good a tentative product is, or to look forward for refining the image. As such, a design process will repeat a series of these three fundamental activities again and again as the operator backtracks to achieve the acceptability (fitness) between the internal coherence (form) and external responsiveness (context) until the entire design problem is solved. The whole design process is a spiral movement (Zeisel, 1981). The notions of internal coherence, external responsiveness, and acceptability relate to the concept of form, context, and fitness

心过程。这心智影像是利用设计者之主观知识为配合某个特定问题而组织出来的方案产品。表现是把建构出的答案化作影像，以各种可能的方法和道具将其外显。至于试验，则是去审核做出的影像（亦即一个设计方案）之可行性，或将设计方案精炼以改进表现。这想像、表现和试验三个主要心智活动就组成了一个设计的过程。

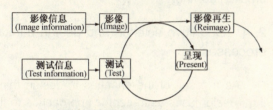

图 2-20 螺旋性重复进行的设计过程
Figure 2-20 Spiral cycles of design processes

根据泽瑟的想法，设计者通常会回溯到过去某段设计的过程再回过头决定当前临时的方案是否够好，或者需要再往前行去改良目前已发展出的影像方案。似此，一个设计过程会以回溯方式重复执行这三个基本活动一直到整个问题解决为止。其目的即在企图达到一个内在凝聚（型）和外界对应（周遭纵横环境）的可接受（妥当）方案。因此这回溯和往前的过程就是螺旋状的运行（Zeisel，1981）。至于这内在凝聚、外界对应和可接受性的企图也与亚历山大所提的型、周遭环境以及妥当性有些相连（Alexander，1964）。但泽瑟对内在凝聚的观念重点是针对型，并且依赖型中每个单元的内部组构和内部单元间是否配合妥当来决定成果。外界对应是指外界周围所有对型的要求。亚历山大认为型是问题的解答，但周围环境的全面背景却定义出问题（Alexander，1964）。泽瑟所提型和环境背景之间的相互可接受的平衡点就正好是亚历山大所提的妥当性之观念（Zeisel，1981）。

从 20 世纪 60 年代到 20 世纪 70 年代，设计方法论的研究开始有一些程度上的进展。研究的成果虽对设计教育有影响，但并不受设计业界重视。阿澈在 1979 年明辨设计师的设计想法有异于科学方法。在科学方法中所用的数字、流程图或逻辑性的模式，不管它们能多正确地描述设计过程中的弹性变动和一些互动性，但毕竟是异类的一种推理方式。在日常生活中，解决设计问题的方法来自人类天性。因此，设计方法必须建立于针对人的基本能力，亦即人类思考的研究。这一番说明强调了对设计思考过程的特性及行为的探索，他认为这些就是设计方法论。亚历山大也对设计方法论开始有不同意见。他在 1962 年发展模式语言的用意，是要创出良好定义的设计程序以便让设计者设计出好设计（Alexander，1984）。但他认为在 20 世纪 70 年代发展出来的设计方法论失败，也不见得对做出好设计有贡献，因此他对方法论的研究方向存疑（Cross，1984）。琼斯和布劳德本特也反对早期的设计方法研究。他们二人认为应该跳出早年系统化的窠臼，另找新路（Broadbent，1979；Cross，1984）。由此开始，设计方法的研究就开始走向分析创作的历程。换言之，由 20 世纪 50 年代到 20 世纪 70 年代，研究是在系统化地描述"什么"是思考的内容。在 20 世纪 70 年代之后，研究重心即转向模索人类是"如何"认知性地进行设计思考和解决设计问题的。

2.4 认知科学与设计思考过程

随着设计思考的改变，结果是设计开始专注以使用者为本位的方法，找出使用者的需求，创造出一些令人满足的解答去解决存在世上的实际问题。因此设计也就是解决问题的一系列心智活动。在 1969 年，司马贺曾在他的《关于人为事物之科学》一书中培育出科学化研究设计的范例，定名为"设计科学"。他指出设计这门科学是一门在知识上深奥，是分析性、半正式化、也半可体验化并可传授的，关于设计过程的学问（Simon，1969）。因此，设计可被理解是一门关于设计过程的一体系知识。

discussed by Alexander (1964). Zeisel's notion of internal coherence addresses to *forms* relying on the inner organization and the internal fitness between the pieces that compose a form. The external responsiveness corresponds to the notion of context which is the part in the world that makes demands of the form. To Alexander (1964), the form is the solution to the problem and the context defines the problem. The mutual acceptability between form and context postulated by Zeisel defines Alexander's notion of fitness (Zeisel, 1981).

From 1960's to 1970's, studies on design methodology had occurred to some extent. Their findings had influenced design education but had not been well received by practitioners. Bruce Archer (1979) argued that designers' ways of thinking are different from scientific ones. In scientific approach, the mathematic, flow-chart, or logical models, however correctly they may be used to describe the flexibility and interactiveness of the design process, are products of an alien mode of reasoning. In everyday life, the ways of solving design problems come from human nature. Thus, design methodology must be based upon the study of fundamental human capacities, which are the ways of human thinking. This argument emphasized the exploration of the characteristics and behavior of design thinking processes, which is, by him, called design methodology. Meanwhile, Alexander had become disillusioned with design methodology. His intention, when developing the pattern language in 1962, was to create well-defined procedures which would enable designers to execute better designs (Alexander, 1984). But, design methodology developed in the 1970's failed to contribute to better design. Thus, he had concerns about the research direction of methodology (Cross, 1984). Jones and Broadbent also rejected much of early work in design methods and were searching for a new approach (Broadbent, 1979; Cross, 1984). Since then, research on design methodology has moved to analyze the creative process. From 1950's to 1970's, research focused on a systematic description of "what" contents of design thinking. From the 1970's onward, some research moved its focus to detect "how" people cognitively process design thinking and problem solving.

2.4 Cognitive science and design thinking processes

As a result of the changes in design thinking, design uses user-centric approaches to find user needs and to create satisfactory solutions for solving existing real world problems. Design is a series of mental activity on problem solving. In 1969, Herbert Simon conceived a design research paradigm titled: "the science of design" in his book "Sciences of Artificial". He indicated that the study of design should be treated as a body of intellectually tough, analytic, partly formalizable, partly empirical, teachable doctrines about the design process (Simon, 1969). Thus, design can be understood as a body of knowledge about the design process, and this body of knowledge can be obtained from empirical observations. The objective of the science of design was to discover the design process by means of which mental phenomena, including human design behavior and cognitive patterns, might systematically be predicted and explained. Since then, research has moved from development of systematic design methodologies for managing the overall design process (Alexander, 1964; Broadbent, 1969; Jones, 1970) to understanding how designers tackle problems with their conventional procedures (Wade, 1977; Cross, 1984; Akin, 1986). This movement reflects the new trend of psychologically modeling the way designers do design.

这种知识可由认真体验中得到,而且设计科学的目标是要发掘设计过程中是哪一种心智现象,包括人类设计行为和认知形态——可被系统化地预测到而且解释出的。从此,研究重心开始从为经营管理整个设计过程而发展系统化的设计方法(Alexander,1964;Broadbent,1969;Jones,1970),转到了解设计师是如何以惯用的程序方法去解决设计问题的新方向(Wade,1977;Cross,1984;Akin,1986)。这一动向显示出一个以心理模式仿真设计师做设计的方法之新趋势。

以模式仿真设计思考的方法之一是认定进行信息处理的特征。关键是在于找出设计者在做设计时,是以哪种制构和运作方式在脑海里处理信息,以及找出适当的表示法用来重现设计观念以便将设计过程模式化。伊士曼(1970)就观察过一件空间规划案的认知过程并解释"产生—并—实验"的过程特色。福斯(1972)也曾观察在设计**原基**(见附记4)的设计行为。埃肯(1978,1979)发展出一组模式解说设计过程中的认知现象并以实验做验证。达克(1979)曾试探在设计时期里,建筑师是否在设计过程中曾经在心脑里发展出顾客的意向或使用者的期待。斯坦尼和马其(1981)也合作以信息处理理论为准,发展出设计语言的观念。埃肯等人(1988)更进一步调研问题构成和问题解决的关系,并说明问题构成可算是一种用来了解设计者如何认出问题因子的方式。陈(1992)也经过分析赖特从1901~1910年期间所做的建筑设计作品和相关著作,研究赖特的设计过程而得出大草原风格的形成理论。大部分这些研究都集中在探讨设计行为,企图找寻出一个正确模式,以便用以解说设计行为所做出的设计产品。

总而言之,从1969年开始兴起的新研究趋势,重心是针对于对待设计活动为解决问题的方式为主。因此,研究的主题开始探讨设计师内心中解决问题所用的心智机制和技巧。然而心智是个黑箱子。心智活动也不是透明可辨的过程,一般很难体会设计观念的出处和创造来源,但这设计迷思已开始受到注意。学术研究也开始探讨创造的过程。但要了解设计的心智过程,理想的研究工具方法在目前而言,就是认知心理学和认知科学。如果设计过程可被明朗化,则设计大师的设计方法可外显,提供公开讨论和学习的机会。设计者也可洞察个人的设计弱点及强处,以增进设计能力。如果设计过程能被程序化,则计算机可做设计,也能增进并促进生产能力。

附记:

1. 韦特墨的视觉组织规律网址:http: //psy. ed. asu. edu/~classics/Wertheimer/Forms/forms. htm
2. 亚历山大的模式语言网址:http: //www. designmatrix. com/pl/anatomy. html
3. 模控学(或译神经机械学)是研究生命体或机械,关于信号(信息)沟通,以及控制流程的学问。
4. 原基这词来自法文,原意是采取立场,建立哲学说理之意。设计原基是为初步设计设立或建造的基本哲学构思,并与最后设计有连带关系。这用法是由巴黎美术学院带头创立,为寻找建筑设计答案而推动出的一种设计方法。

One approach to modeling design thinking is to identify the characteristics of humans processing information. It is critical to recognize mechanisms and operators that designers used to process information in the mind while they were working on design, and to have appropriate conceptual representations that can be used to model the process. Eastman (1970) observed the cognitive processes in a space-planning task and explained the characteristics of "generate-and-test" processes. Foz(1972) observed design behavior in a **parti** (see note 4) design. Akin (1978, 1979) developed a number of processing models to describe the cognitive aspects that occurred in design processes and justified them with experiments. Darke (1979) intended to find whether architects had in mind an image or expectations about users during their design period. Stiny and March (1981) developed notions of design languages starting from the information processing theory. Akin et al. (1988) further investigated the correlation between problem structuring and problem solving, and reported that problem structuring could serve as a mechanism to see how problem parameters were identified by a designer during a design process. Chan (1992) studied Frank Lloyd Wright's design process from observing Wright's works completed from 1901 to 1910 and theorized the resulting Prairie Houses style created by Wright. Most studies have concentrated on exploring the design actions and finding an accurate model that explains the design produced by the design actions.

In summary, the new research movement was centered on treating design activities as ways of solving problems. Thus, research concentrations were on exploring designers' cognitive mechanisms and mental techniques on problem solving. The human mind, however, is a black box and the mental processes are not transparent. Thus, it is difficult to comprehend design origins and sources of creativity. But, the design myth has been noticed and research works have begun to explore the creation processes. The most appropriate research tools, in this regard, are cognitive psychology and cognitive science. If the design processes could be explored, then master designers' design processes could be explored, discussed, and learned. Individual designers could discover their own design weakness and strength for improving design ability. If the design processes could be programmed, then computers could do design, which would improve the capacity of productivity.

Note:

1. URL of Wertheimer's organizational law Web page: http://psy.ed.asu.edu/~classics/Wertheimer/Forms/forms.htm
2. URL of Alexander's pattern language Web page: http://www.designmatrix.com/pl/anatomy.html
3. Cybernetics is the study on information communication and sequences control.
4. Design parti in France means to take a stand, establish a philosophical position. The design parti forms the basis of the preliminary design philosophy relative to the final solution. The use of parti has been promoted by the Ecole des Beaux-Arts school in resolution of architectural design problems.

第3章 认知的组构要素

人类认知是经由感觉器官去吸取信息，再由一些认知组构单元去了解、处理、储存、回收，然后利用得出的信息结果解决现实问题的一系列心理过程。在这些过程中，所有的认知活动都有一些认知组构单元参与。有些认知组构单元主宰了人的感觉系统，控制了过程，也左右了其他相关活动。但在某些情况及处境下，信息处理过程会产生不同的信息运作方法，而产生不同的效果。这些涉及的认知心理活动相当复杂，但有能力把认知过程予以科学法定位，并透视解说认知运作的方法之一是**人类信息处理学**。本章将把信息处理学作一说明，将认知活动中专职负责的认知组构单元和人类思考系统中占有主要地位的认知因素作一叙述。相信在了解认知组构单元之后，能对认知这一领域有充分信心，对设计思考的心理活动开始有明确的掌握，对设计的特性作科学化的解释，更进一步地也可清楚地认定思考逻辑，并衡量设计条理。

3.1 信息处理理论

心智活动具有一系列可区分的活动片断，信息处理理论这学科即试着去辨证每个时期发生的事情。这领域的学者也相信思考就是在脑海中做信息处理及加工。因此信息处理理论的重点是研究在心智中的信息是如何被收录、储存和运作的。细节如下：

（1）信息的收录：涉及**注意力**、**知觉**及**辨识**。注意力是主要控制适当信息收受的过滤器。知觉是把感觉收到的刺激转化为有组织的心理体验之能力，或人类诠叙感觉到的信息去体会这世上初步并直接经验的功能。辨认是把收入信息作一区分认证。这些心智活动都曾经过心理实验，证明它们的存在。

（2）信息储存：涉及三个储存区，即**感觉收录器**（或感觉缓冲储存区）、**短期记忆**和**长期记忆区**。信息处理理论也十分专注于知识的表征（或重现表示），探讨知识以何种式样存放于记忆中。

（3）信息的运作：是更进一步的认知进行，包括数据搜寻及信息分类。由于信息来自看、听、触、嗅、尝等五种感觉。其来源庞大且交互重叠，所以有必要以系统化的方法去明确说明记忆中的数据是如何被储存和搜寻的。

基本而言，人类从环境中经由"感觉收录器"（即感官）接收信息。这感觉收录器是心智进行运作的单元之一，它将信息接收后，会将信息以收到的感觉原形作短期暂时存放（Sperling，1963；Atkinson & Shiffrin，1971）。每个人类感觉神经都有其感觉收录器。但视觉（Sperling，1960）和听觉（Neisser，1967）两个收录器是目前最被广为研究的（见附记1）。例如，研究显示视觉收录器保存信息的储存期（Sperling，1960）是1/4秒（250毫秒），但保留的时

Chapter 3　Cognitive mechanisms

Human cognition is a series of mental activities that gather information by sensory systems; understand, process, store, and recycle information through certain cognitive mechanisms; and finally apply the resulting information to solve physical problems. During the process, all activities have certain cognitive mechanisms involved. Some components dominate our sensory system, control the process, and affect other related activities. The processing of information under certain situations and conditions occurs differently and yields different results. The involved psychological activities are very complicated and subtle. One of the approaches to scientifically identify the process and explain the operation is **human information processing theory**. This chapter explains the information processing theory, introduces the major cognitive mechanisms, and key cognitive factors existing in thinking. It is believed that understanding these cognitive components will provide more resources to better handle the cognitive activities occurring in the design thinking processes, to scientifically explain the characteristics of design, and to clearly identify thinking logic and design rationales.

3.1　Information processing theory

Mental activities have a series of separate stages. Information processing theory attempts to identify what happens during these stages. Scholars in this area believe that thinking is information processing in the mind. Thus, studies in information processing theory concentrate on how information is (1) received, (2) stored, and (3) operated in the mind. Details cover the following:

(1) Information input relates to **attention**, **perception**, and **recognition**. Attention is the major filter that controls the appropriate information input. Perception is the faculty that converts the presented sensory stimuli to organize psychological apprehension, or the function of interpreting sensory information to get immediate and direct experiences of the world. Recognition distinguishes the received information. The existences of these mental activities have been proven through psychological experiments.

(2) Information storage relates to three areas of **sensory register** (or sensory buffer), **short-term memory** (STM), and **long-term memory** (LTM). The information processing

间通常会长到信息被辨认后为止（毫秒是微单元，1秒等于1000毫秒）。然后信息就会被移到短期记忆中作处理，处理结果会决定信息是要被删除或是转放到长期记忆中。信息的进行如果是由吸取、辨认到储存的方向依序推进，则称为由下往上或**数据导向**的进行方式。如果是相反地由记忆抽取，经辨认到应用，则称为是由上往下或**观念导向**（Lindsay&Norman，1977）的进行程序。这些信息处理学科发展出的理论，已被广泛地运用于人工智能、学习及语言学中作为仿真人类思考的基本架构。

3.2 知觉与注意力

感官是信息接受器。听与看是重要的感觉器官，因为在工业社会中与人切身相关的信息相当浩巨。适当的信息必得先被看到或听到，才能立刻对外界产生反应。也如此，眼睛被隐喻为灵魂之窗，耳朵被称为是声音的收录器。通过这窗口和收录器，人类可知宇宙世界中呈现的情报，捕捉到情报中的信息，演绎信息中的含义，然后把了解的含义组织起来放在记忆中。这整个过程，即是**知觉**的现象。简言之，知觉即是演绎、诠述、了解知觉信息，把由收录器接受到的对象或事件附加以意义的过程。即使知觉是由不同的感官途径中经过收录、转译、选取、安排并组织信息的一系列综合认知活动，但注意力是这些系列处理信息活动的主导。事实上，注意力不只是决定什么情报被注意到，被演绎出；并且也影响是何种情报被选取，被进一步储存的一项主要认知组构要素。

注意力是导引心智在某一时刻集中于单一事件或单一刺激的心智效果。注意力的特征有二，第一是专注性。专注性是多关注并多付心智精神去捕抓目标信息，有名的例子是**鸡尾酒会现象**（Cherry，1953）。如果请客的主人用的是和客人相同的本土母语，而且周围没有其他会话发生，则不必太费劲就可了解主人的对话内容。如果周围有一些其他对话，那么客人就要花些工夫去配合维持会话速度。如果主人用的是不同的外地语言，而且周围还有一些其他活动存在，那么客人就需要更专心集中于主人的对话才能捕捉情况。如果客人在与主人对话时还想分心聆听其他客人的对话，则需要付出更多的精神和心智努力才能达到旁听的目的。

第二个注意力的特征是选择性。选择性是有选择地注意周遭存在的信息。这个选择的本能会让人减轻情报的负荷量，包括少花时间收录每个信息，忽视低重要性的信息或完全排除收录。最早的选择理论是瓶颈理论。布劳德本特（1957）指出注意力的功能像是一个过滤器，每次发生知觉认知的时刻，这过滤器只处理单项信息。瓶颈的观念是个象征性Y形管状的过滤模式。Y形顶部两个分叉代表两个耳朵，Y形下部是个狭窄管，每次只让一个信息通过。在分叉和狭窄交合处是一个活动过滤板。其功能是允许信息由两个分叉往下流入狭窄管。如果两个信息同时发生在两个耳朵时，这过滤板会挡住一边耳朵让另一边耳朵的信息流入，同时把挡住的耳朵之信息暂时存放在感觉储存器中。如果听者被要求回报说出发生在左耳的三个信息，则这过滤板会挡住右耳，让左耳三个信息完全通过之后才会换边。如果听者被要求回报给实验者，说出两个耳朵同时发生的所有信息，那么这活动板就得快速地来回换边，让全部信息都能流通进入。这观念假设听者每次只能选择性地辨认一个管道中的情报。

相反于信息每次只能在一个管道中辨认的观念，马里（1959）发现受测者有时在没有注意的听觉管道中也会听到他们自己的名字。特瑞斯曼（1960）也发现有些字有低于其他字的听觉门坎，以利辨认。比方说，受测者的姓氏和一些较有危险意义的信息，如"火灾"等。这些模式建议某些字有较低的辨认频率，也容易在没被注意的管道中被分辨出。道曲与道曲（1963）的实验即在两只耳朵中听两个不同会话，所有的字在专注的管道中被听到的是重要的，

theory also concentrates on knowledge **representation**, how knowledge is stored in memory.

(3) Information operation relates to advanced cognitive processes which include searching and categorizing. Human information comes from five sensory inputs: visual, audio, touch, smell, and taste. With vast resources overlapping, it is necessary to have systematic methods to explain clearly how information in memory can be stored and searched.

Basically, human beings get information from the environment through their "sensory register", which is the mental processing unit that receives information and holds the information temporarily in its original sensory form (Sperling, 1963; Atkinson and Shiffrin, 1971). A register presumably exists for each of the senses: visual (Sperling, 1960) and auditory (Neisser, 1967) registers have been widely studied (see note 1). It is reported (Sperling, 1960) that the visual register can hold information for about one quarter of a second (250 msec). But the time can be extended until the information is recognized. (Millisecond is a micro unit, one second equals 1000 milliseconds.) Then the information is moved to short-term memory for processing. Results of the processing will determine whether to erase the information or move it to the long-term memory for storage. If the information movement is from input, recognition, to storage, it is called bottom-up or **data-driven process**. If the information comes from memory through recognition for application, it is called top-down or **conceptual-driven process** (Lindsay & Norman, 1977). These theories have been applied in fields of artificial intelligence, learning, and linguistics to simulate human thinking.

3.2 Perception and attention

Sensors are information receivers. Visual and auditory sensors are two important ones. In industrial society, there are tremendous information resources that are critical and should be seen and heard first in order to react. Eyes are metaphorically named the windows of the mind; ears are receiver of sound. Through the window openings and sound receivers, human beings can perceive a variety of information in the world, catch the contained message, interpret the embedded meaning, organize the comprehended meaning, and save it into memory. The entire process is the phenomenon of **perception**. In short, perception is the process of interpreting and understanding sensory information by attaching meaning to objects and events sensed by the receptors. Even though perception is the comprehensive mental activities of receiving, translating, choosing, and arranging sensory information obtained from various sensory channels, attention also plays an important role in guiding these information processing activities. In fact, attention is a key cognitive mechanism that determines not only what information is focused on, interpreted, and recognized, but it also affects which information is collected and stored.

Attention is a mental effort that directs the mind to concentrate on one task or one stimulus at a time. There are two characteristics of attention. The first characteristic is concentration. Concentration means to attend to and focus mental efforts on capturing target information. The famous example is the **cocktail party phenomenon** (Cherry, 1953). If the party host speaks the same native language as the guest does and there is no other conversation around, it will take little effort for the guest to understand what the host is saying. If there are other conversations occurring in the surroundings, it will take more effort to maintain the flow of the conversation. If there are activities around and the host speaks a different lan-

因为听者必须要"跟读"。字在不被专注的管道中听到的是不重要的，也会被忘记的，因为听者在这实验中被要求专注于另一管道。这结果显示在两个管道中听到的两个会话都会被辨认，但很快即会被忘记，除非这些会话都是重要的。这说明辨认信息不但是因门坎高低而定，同时也取决于输入的重要性。重要与否，即由注意力判断。如果听到的字是不重要的，则很快即被遗忘。但如果所有的字都是重要的，那么在执行一事情时所需要付出的心智努力就涉及另一项重要的观念，即能量理论。

能量理论假设每人都有某些执行心智活动的能量限制额，也有对不同活动斟酌赋予不同能量的控制。某些活动需要比其他活动付出更多的注意力能量。如果对某个活动所赋予的注意力能量不足，则其执行效果和表现质量会减弱（Kahneman, 1973）。例如，设计师做设计时可同时收听广播，这是假设在做这两件活动时付出的注意力能量并不超越所需的能量限制额。但如果设计问题是个难题，则设计师不易在做设计同时，也能充分掌握由听广播中得到的信息。但设计者可在危急时刻二者取一，选择专心听广播或专心做设计。这能量理论指出干扰会在活动超出能量时发生，但瓶颈理论指出是因一个同样的认知制构被同时用到两件不匹配的执行上，所以干扰才会发生。

瓶颈和能量理论都说明了注意力是个有限的心智资源，也说明了其选择性之本质。选择性的发生时间会在接受信息的早期，又称为"**输入注意力**"，用以处理属于较低层次的信息，而且处理的时间非常快速。比方在视觉信息的处理中，视者必须在短暂视觉展示中决定哪些信息可以被抽选。史朴令（1963）即做了实验给视者快速地展示三列四个字母影像约 50 毫秒时间。这影像关闭之后，视者会听到声音信号，信号有高、中、低三音代表影像的上、中、下三列。视者即根据声音信号回报视觉展示中该列的影像。这种方法是"**部分回报程序法**"。

史朴令所用的部分回报程序法可根据在影像关闭时视者已经记住的字母，来确定有多少字母被视者注意到并记存住。结果证明视者可回忆一列里大约多于三个字母。因为视者事先不知要回报那一列，他们得观察每列多于三个字母以备回报。这说明每次回报时，大约总共有多于九个字母会暂存于视觉储存区作预备。此法异于当时 1960 年代使用的"**全体回报法**"。全体回报法是要视者回报每次展示中全部看到的字母。实验结果显示，不管展出的字母有多少，每次展示的回报率大约是在四到五个之间。史朴令也把信号时间作长短安排，即视者等信号结束后再回报，结果也证明信号延长时间越长，回报的字母越少。如果信号长到一秒，视者就跟全部回报法一样，只能回报四到五个。因此，所有被注意到的数据是短期地被存在视觉储存区里待进一步的心智进行。停留时间过长，没有被处理的数据都会逐渐消失。这说明了信息自外界输入之后被暂时存放在视觉收录器里的现象。但并非所有事件都得经过这辨认程序，有些事件很快，有些更是自动化地被辨认。

自动化的程序出现在处理一些惯常发生的事件中。哈瑟和扎克（1979）即证明了某些事件经过充分练习之后，就会自动化。这说明了一些信息被学到之后，即存在记忆里，而后不费劲地即可自动使用出来。导致"自动处理"的事件极多是处理时常发生、有形态，而且日常例行的信息情报。骑车即是一例，开始时要花很多注意力专心学习如何骑车。当骑熟之后，即可一边骑车，一边作其他思考。因为这时技术已熟练，不必再花心思注意踩踏。阅读是另一例，学习阅读（LaBerge & Samuels, 1974）开始于分析字母的特征，组合这些特征去认清字母，组合字母去认清单词，清楚每个单词的意义，然后组合这些字的意义去了解整个章节含义。经过纯熟练习之后，字母特征可以很容易地被组合成单元并轻易地被辨认。乐部格和塞缪尔指出自动认出字母

guage, even more concentration will be required to maintain the conversation. If the guest wants to listen to one of the other conversations while listening to the host, more concentration and mental effort are needed to absorb the other conversation.

The second characteristic of attention is selectivity. Selectivity means to selectively pay attention to the information presented in the surroundings. This activity helps humans reduce information overload, by spending less time on each input, disregarding low-priority inputs, or completely blocking sensory inputs. The first theory of selectivity is bottleneck theory developed by Broadbent (1957). He specified that attention acts as a filter which processes only one message at a time at the perception stage. The bottleneck concept is a symbolic filter model that has a Y-shaped tube. The top two branches of Y symbolize the two ears. The lower Y is a narrow stem that accepts only one signal at a time. At the junction of the branches and stem is a flap filter accepting the flow of signals that enter the stem. If signals occur on two branches, the flap filter would be set to one side to allow one signal entering the stem, while the other signal would be kept in a sensory store. If a listener was required to report three signals occurring in the left ear, the flap filter would stay on the right ear until all signals from the left side entered. If the listener was required to report all signals to the experimenter when they arrived on both sides, the flap filter would shift rapidly to allow all signals to enter. This assumes that a listener can recognize information selectively on only one channel at a time.

Contrary to the concept of recognizing information on only one channel at a time, Moray (1959) found that subjects sometimes heard their own names on the unattended channel. Treisman (1960) also discovered that some words have a lower threshold than others for easy recognition; for instance, the subject's personal name and some danger signals such as "fire". These models suggest that some words have low thresholds that can be heard on unattended channels. Deutsch and Deutsch (1963) studied hearing two different conversations on two channels (ears). Words heard on the attended channel are important so the listener must "shadow" them. Words on the unattended channel are usually unimportant for that the listener is asked in the experiment to attend to another channel. Results show that words in both conversations are recognized and quickly forgotten unless they are important. This implies that information is selected not only by the threshold level, but also by the importance of the input which is judged by attention. If words are not important, they are quickly forgotten. If words are all important, then it requires more mental effort to perform a task, which relates to the capacity theory.

Capacity theory assumes that a person has a certain limit for performing mental tasks and has control of the capacity that can be allocated to different activities. Some activities require more effort than others. When the attention allocated to the activity is not enough, the execution effort and performance quality decline (Kahneman, 1973). For instance, a designer can render a design and listen to a radio at the same time if both activities do not exceed the capacity for executing the two different tasks. However, when the design problem is a difficult one, the incoming message from the radio will not be able to be fully comprehended while working on the design problem. But the designer could choose only to listen to the radio or only to work on the design, instead of on both, at a critical moment. This capacity theory indicates that interference happens when the activities ex-

的过程会让人认得字母是总合的特征单元,而非各自的特征。这会节省用在花耗注意力的能量,并可将省下的能量转用在其他所需技能上。赫利(1980)也解释读者习惯于将常发生的词当作是单元来解而较少注意细部的字母特征。因此,常用的字是熟悉的组合体块,容易辨认,也让人少花能量和精神去注意各个字母。

3.3 形态辨认

很清楚地,基本上一些信息是被注意到才被收录的,而且先被记录在感觉储存区里,然后再一次经由注意力作进一步的运行。在这时刻,非常重要的课题是要了解人类怎么辨认知会这些信息情报。这涉及认知技巧中的形态辨认,也牵涉注意力的运作。形态辨认中有三个在认知学科中已被广为探讨的重要理论,即样板比对、特征分析和结构描述。

样板比对理论指出一个被感受到的影像会被编码到感觉储存区里,以备与记忆中已储存的许多形态作比对。这些已储存的形态被称为样板,也有**全体宏观**的特质(见附记2)。样板理论的基本观念是说被编了码的影像如能与记忆中存有的样板相配对合,则这影像就被认同。然而,这样板比对理论有下列的问题存在,即这编码信号在还没被认出之前,是如何得以被保留在感觉储存区里的?是保留先,还是辨认先?换言之,在还没与样板配对之前,这没被认出的信息影像,是怎样能被正确地注意到,而且无误地依所需要的原形依样存放在感觉储存区里的?

由于样板比对理论存在着上述的困难,特征分析理论就被提出。特征分析理论是说形态辨认发生是着力在分析特征。于辨认影像时,人类会分列部分特色来形容形态。吉布森(1969)提议说视觉学习的发生是由认证特征来区分不同的形态。在他的实验中,他提议发展出一整套完好可靠的特征组构来预测视觉混淆的可能性。因为容易混淆的特征群一定是有许多相同的特征存在。他的经由在视觉上比较异同特征作分析的方法(Gibson, Schapiro & Yonas, 1968),也更进一步的被汤绅(1971)运用在测量特征与特征间的相似性上。

结构理论强调特征间的关系,要求设立详细规格说明所有特征是如何配置的。类似于完形心理学,这结构理论会明列特征如何合并成形态或形态构件,这也就是"详细规格"之所以说明了形态与形态间的结构框架。结构理论的研究重点专注于字母和数字影像是如何被辨认的。由于结构理论例证了线条是如何装配起来的,因此它比特征分析学说还要更清楚地解释出线条的形态是如何被辨认的。例如帕麦尔(1977)就用实验证明形态的辨认一方面可以把形态分解作为单体作分析辨认,另一方面也可以把形态组合作综合归纳辨认。

帕麦尔的实验是把形态(图3-1上方两个图形)分割成高质、中质、低质的系列个体(图3-1下方12个图形)。图3-1上方两图是由下方成对的部分组合而成。理论上,构件的结构可决定心智重组的速度。好的个体(图3-1的H即高质部分)比不自然的个体(图3-1的L即低质部分)较易于合并。实验过程是受测者看了两个个体之后,被要求在心智上把这些个体重合成形态。形态完成即按电钮,此时一个完全正确的或是相似的形态结果会显示出来,受测者就要回答这影像是否就是他们刚完成的合成结果。实验结果显示,受测者花一秒半时间完成高质组件,四秒半完成中质或低质组件。这发现支持在组配形态时有可能是心智中已有预先定好的组合方式存在。这也说明了辨认熟悉字母是自动化的程序而不需要注意力。但在辨认不熟悉形态时,注意力是必须要付出才能把特征组合成形态的。注意力在这阶段扮演了重要的角色。

形态辨认主要研究特征是如何被认出的。前面所提的样板比对、特征分析和结构描叙理论与由下往上的分析过程有关。因为信息在人脑中流动的过

ceed the capacity, whereas the bottleneck theory indicates that interference occurs because the same mechanism is required to carry out two incompatible operations.

Both bottleneck and capacity theories have explained that our mental attention is limited and selective in nature. The timing of selection occurs at an early mental processing stage, called **input attention**, handling low level information with rapid speed. For instance, in visual–information processing, viewers determine what information can be extracted from a brief visual presentation. Sperling (1963) presented three rows of four letters for a brief exposure of 50 msec. After the stimulus was turned off, viewers were cued; which had three different tones, with high tone for the top row, medium for middle, and low for bottom, to report just that row of the display. This method was called "**partial–report procedure method**".

The partial–report procedure method used by Sperling was based on the number of letters that a viewer could recall to determine the number of letters retained and stored when the display was turned off. Results showed that viewers could recall over three items of letters from a row. Since viewers did not know beforehand on which row of the display to report back to the experimenter, they had to have more than three items per row for preparation. Thus, there are more than nine items stored and available in the visual sensory ready for report. This method differs from the "**whole–report procedure method**" used in the 1960's. The whole report procedure method asked viewers to report as many items as they could recall from one display. Results showed that many viewers could report four to five items of letters regardless of how many letters were presented. Sperling also made variations on the cue time. Viewers waited until the cue tone ended before reporting. Results showed that the longer delay time of the cue, the less number of reported letters. When the delay increased to one second, viewers' performance were the same level as the whole report method of four to five items per display. Thus, the visual sensory store holds temporarily all the attended information for further processing. If the information had been held longer in the register without processing, more items would have been lost over time. This explains the phenomenon that input information is held temporarily in the visual sensory register for recognition. But, not all the things or items need to go through similar procedures of recognition; some can be processed quickly, even automatically recognized.

Automatic procedures occur in handling some routine events. Hasher and Zacks (1979) demonstrated that some tasks with sufficient practice can become automatic. This explains that some information was learned, stored in memory, and later used automatically without much mental efforts. Tasks that triggered automatic processing were frequently occurred, patterned, and routine information. Bicycle riding is an example. It requires considerable attention to learn to ride. After the skill is mastered, one could do other thinking and need not concentrate on the riding while bicycling. Reading (LaBerge & Samuels, 1974) is another example. Learning to read begins with analyzing the features of letters, combining the features to recognize the letters, combining the letters to recognize words, understanding the meaning of each word, and combining the meaning of the words to comprehend the meaning of the text. After sufficient practice, features can be combined to form letters and recognized easily. LeBerge and Samuels indicated that the automatic processing of perceiving letters let people recognize letters as units instead of individual fea-

程缘起于感受到的各片特征，这些特征是配置成可辨认大形态的认知基础。而大的形态认知就是由这些小特征建构而成的。然而，在视觉与听觉的信息输入中，一般的背景脉络情报也会引导认知（Warren & Warren，1970）。这现象就涉及由上往下（**观念导向**）的分析过程，在这过程中，高层次的知识决定了如何演绎低层次单元。例如阅读，由下往上（**数据导向**）的分析过程始于认出我们所看到的字母特征。而相反的，由上往下的分析过程则来自使用语言知识去做领悟工作。这语言知识即包括拼字、字词、语法及语义知识。

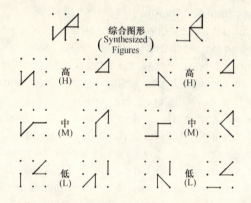

图 3-1 与心智综合实验中用的形态
Figure 3-1 Patterns used in a mental synthesis experiment（Reprinted from "Hierarchical Structure in Perceptual Representation", by S. E. Palmer, *Cognitive Psychology*, 9, p. 463, 1977, with permission from Elsevier）

拼字知识指的是不同字母组成字符串的可能性。例如单词中 U 通常跟着 Q。字词知识涉及字母的组合形成一个语言中的单词。语法知识指的是语言中的文法，语义知识涉及阅读到的字词之意。在阅读过程中，读者会用上下文背景脉络去帮助演绎出信息，也会用推理、利用语言知识去推导认不出的信息。比方说在看笔写的书信时，可能有些笔迹暧昧无法辨认，但由上下文脉可推理出可能的用字，

也有可能有时字词有误用，但字里行间也可提供信息猜得出所具备的大意。因此，读者会同时辨认拼字、拼音及单字（这是由下往上分析法），同时确定附带的意义（这是由上往下分析法）。这两种方法都必须得要由某种途径合并使用才能使认知过程发生。有技巧的读者通常都精于两法。

信息的前后背景脉络会被用来作附加值，补充形态辨认，以利认知感觉，这也是影响感知的因素。比方说脸孔是耳、鼻、眼和唇的综合。这些整体组合依上下脉络而定，就可被认出是一张特殊的脸孔。经由这些脉络，不必太花工夫去辨认各单元结构，如鼻或唇，这张脸孔即可被认出。但如果这些单元部分被隔离展示，失去脉络的支持，则需要更多数据（Palmer，1975）才能认出这些个别的鼻或唇。这解释了背景脉络在视觉认知感上扮演着一个重要的角色。同样，在语音辨认上也是同样重要。例如华伦（1970）的实验证明了关键词由跟在它后面的词来决定。米勒和依莎德（1963）也证明有英语文法的句子比没有英语文法的句子更易辨认。因此，如果听者没听到句子中的一个单词，但随后的词和整个脉络可提供更多的资料去补充了解整个句子的含义。

背景脉络同时也影响视觉认知，例如图 3-2（a），所见到的是垂直五行圆圈，而不是水平五列圆圈。因为柱间圆圈距离比列间还接近。这图例说明完形心理学中的相近律解释了脉络对认辨形态的影响。完形心理学中讨论的对称律也决定了形态是如何

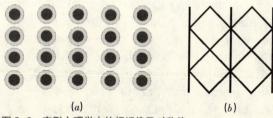

图 3-2 完形心理学中的相近律及对称律
(a) 相近律；(b) 对称律
Figure 3-2 Illustration of the Gestalt principle of proximity and symmetry
(a) proximity; (b) symmetry

tures which free capacity for other necessary component skills. Healy (1980) also explained that readers get used to frequently occurring words as units and rarely focus on individual letters. Therefore, the frequently used words are familiar units that are easily to be recognized, requiring less effort and attention to the individual letters.

3.3 Pattern recognition

It is clear that fundamentally some information is received by attention and recorded in the sensory storage first. Afterwards, information is again selected by attention mechanisms for processing. However, it is critical at this point to understand how human beings recognize this information. This relates to the cognitive skill of pattern recognition, which also involves the operation of attention. Three important theories related to pattern recognition—template matching, feature analysis, and structural descriptions—have been extensively explored in the field.

Template matching theory indicates that a perceived image encoded in the register storage area will be used to compare various patterns stored in memory. These stored patterns are called templates and are **holistic** (see note 2) in nature. The basic concept is that as long as the encoded image matches the template in memory, it is recognized. This template matching theory has been challenged by asking the questions: "How is the encoded information preserved in the sensory store if it is unrecognized? Which goes first in the process, the recognition or preservation?" In other words, how can unrecognized information be attended precisely and encoded accurately before it is matched with the template in memory?

Because of the difficulties on template matching, feature analysis theories propose that pattern recognition happens through analyzing features. While recognizing images, humans describe patterns by listing the parts. Gibson (1969) proposed that learning through perception occurs when exploring features that distinguish one pattern from another. In his experiments, Gibson proposed to develop a reliable set of features for predicting perceptual confusions, because confusable items share many common features. His method of feature analysis through perception by comparing likeness and differences among features (Gibson, Schapiro & Yonas, 1968) was further studied by Townsend (1971) when measuring similarities between features.

Structural theories emphasize the relationship among features. It involves specifying through definition how the features fit together. Similar to Gestalt psychology, structural theory specifies how features are joined together into patterns and parts, which is the "specification" of the structure of patterns. These studies emphasized how the images of letters and numbers are recognized. Since the theories illustrate how lines are joined together, they give a better explanation of line patterns than feature analysis models can do. Palmer (1977) demonstrated that pattern recognition can occur by breaking a pattern into parts as the tasks of analysis, or by combining parts to make a pattern which is the task of synthesis.

In Palmer's experiments, patterns (the 2 figures in Figure 3-1) are divided into high, medium, and low goodness parts (the lower 12 figures in Figure 3-1). At the top of the Figure 3-1 there are two figures that are synthesized by combining the pairs of parts shown below them. Theoretically, the structure of the parts could determine the mental speed of combining parts. Good parts should be easier to combine than unnatural parts. Subjects were shown parts and were asked to mentally com-

被认知的。图 3-2（b）中的形态会被看成是三根垂直线和五个钻石形，而非几个正放及颠倒反放的大 K，因为钻石是对称形，而 K 非对称。非对称不易被知觉认出。因此不同的背景脉络安排及形态特性会产生不同的视觉感受。相关的影响已在完形心理学中被充分研究探讨了（见第 2 章完形心理学介绍）。

3.4 心智影像

当人看见一个物体时，这物体的原始形状或形态就被铭记在感觉储存区里。在形态辨认期间，这物体就经由辨识其潜存的特征而被认出形态。形态被认出之后，知觉信号会被进一步地处理、诠述、组织，并存放于记忆中（kosslyn & Pomerantz，1977）。有趣的是，这些被认出的信息是如何存放在记忆中，以及记忆中存放的信号是以哪种形式存放的？这就全部涉及心智影像或视觉影像这一研究领域。

心智影像（或心像），如学者提议的，是一种空间表示法，和视觉感知看到物体时的视觉经历相似。其表示法具备有大小、尺度、色彩和某种程度的质感，也代表真实的物体。因此，当物体被看到之后，一个影像图片就在心中成立，这在心中看到的图片就如真实看到的物体一般。这现象与心像扫描的观念有关，即在心中重建的心像图片与实际视觉影像类似。这心图和实际视觉影像的相似性可由扫描时间和实际距离建议出（Kosslyn, Ball & Reiser，1978），即从扫描完一个物体转移到另一个物体的时间与二者间的距离有关。在他们的实验中，柯斯林和他的同事使用一个假想的海岛图，图上有一些人造物，包括一颗树、石、茅屋、井、湖、沙丘和绿地（图 3-3）。受测者被要求仔细记住这地图细部，然后把这图的影像在心中重现。尔后一个物体的名称被大声念出。受测者要在心中扫描专心定位这念出的物体。四秒钟后，第二个物体被念出，此时受测者被要求要扫描、寻找这第二件物体，并决定它是否存在于这地图上，其过程是想像有一个黑斑点从第一个记住的物体移动到被念到的第二个物

体。当受测者找到并且到达这第二个物体时，即按"到达钮"把定时器停止。这定时器是用来测量**反应时间**的。如果在图上找不到则按另一钮。结果显示物体越远，所需反应时间越长。换言之，如实际物体的距离增加，心像扫描时间也增长（图 3-4）。如实际物体的空间关系是非直线时，心像扫描也是非线性的。

图 3-3 测试心智扫描所用的岛图
Figure 3-3 Maps of an island used to test mental scanning (Reprinted from "Human Perception and Performance", by S. M. Kosslyn, T. M. Ball, and B. J. Reiser, *Journal of Experimental Psychology*, 4, p. 51, 1978, with permission from American Psychological Association)

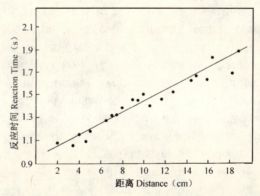

图 3-4 扫描时间与图 3-3 中所有二点距离之间的函数
Figure 3-4 Time to scan between two points in Figure 3-3 as a function of the distance between points (Reprinted from "Human Perception and Performance", by S. M. Kosslyn, T. M. Ball, and B. J. Reiser, *Journal of Experimental Psychology*, 4, p. 52, 1978, with permission from American Psychological Association)

bine the parts to form patterns. After the synthesized pattern was completed, subjects pressed a button, a pattern either similar or identical to the correct synthesis was presented and subjects responded whether the pattern matched their synthesized pattern. Results showed that subjects took about 1.5 seconds to mentally construct a pattern from high-goodness parts and around 4.5 seconds to construct a pattern from medium or low-goodness parts. The findings supported the theory that there is a predominant way of organizing the features of a pattern. It seems, unlike the little attention needed to recognize familiar letters, which is done automatically, recognizing unfamiliar patterns requires attention to synthesize the features into a pattern. At this stage, attention plays a critical role.

The major study of pattern recognition focuses on how features can be recognized. The theories of template match, feature analysis, and structural descriptions relate to the bottom-up analysis process. Because information flows in our mind from perceived pieces of features, that serve as the foundation of perception, to larger units built from them. However, in visual and audio input, the general context also guides perception (Warren & Warren, 1970), which is referred to as the top-down (**conceptual-driven**) analysis process. The high level knowledge determines the interpretation of low-level units. For instance, in reading, the bottom-up (**data-driven**) analysis begins with the features that we are seeing. The top-down analysis begins with the use of our knowledge of language for comprehension, which includes orthographic, lexical, syntactic, and semantic knowledge.

Orthographic knowledge relates to probability in various strings of letters. For example, a U typically follows a Q. Lexical knowledge refers to the letter combinations that form a word in a language. Syntactic knowledge relates to the language's grammar, and semantic knowledge is concerned with the meaning of the words we read. In the reading processes, context is used to aid in interpreting perceived information, and inferences are made from the knowledge of language to convey information that was not perceived. For instance, it might be difficult to recognize all words when reading handwriting. We might see a letter or word as ambiguous or illegible. But, the context makes it clear which word is intended. If a word is misused, the syntax could provide information on the intended meaning. Therefore, we recognize alphanumeric characters and words, which is bottom up analysis; and simultaneously determine their associated meanings, which is top-down analysis. The two approaches must combine in some way for the perception process to occur. Skillful readers are always good in both.

The context in which information is used is a supplement to help pattern recognition, and is the factor that affects perception. For instance, a face is a combination of ears, nose, eyes, and lips. The combination of these elements can be recognized through its context as a particular face. Through the context of a face, little effort is needed to recognize its individual parts, for instance, the nose or lips, for recognizing the face. Yet, when these parts are presented in isolation, more information is needed for individual recognition (Palmer, 1975). This explains the important role that context plays in visual perception. Similarly, it is important in speech recognition. For instance, Warren (1970) demonstrated a critical word is determined by subsequent words. Miller and Isard (1963) demonstrated that sentences obeying the rules of English grammar are easier to understand than ungrammatical sentences. Furthermore, if a listener misses hearing a word in a sentence,

心像运作更可以是三度空间运作。例如，谢巴德和麦兹勒（1971）即用三维物体的二维图像去探测人类的视感。在测验物图形中，一对对物体由不同角度在图形上显示（图3-5）。受测者必须决定这对物体是否为相同物体，而非其方位角度是否相同。例如，图3-5（a）所示的一对物体是相同物体，但不同之处是右边物体沿图片面（即 XY 面或 Z 轴，见附记3）沿顺时针方向转80°。图3-5（b）所示的一对也是相同物体，但右物在深度面（即 XZ 面或 Y 轴，见附记3）顺时针转80°。在实验刺激中，所有成对物的不同方位角度差距，是以20°为间隔，差距幅度自0°到180°。所有由计算机产生之1600对测验物里，其中一半（800对）可以转动到完全相配，另一半（800对）是镜像，并不相配（图3-5c）。图3-5（a）、（b）仅是所用许多可相配实验物中之两例。

实验结果显示转移的方位角度之增加度和需要的反应时间之增加度为相等值。物体之间的转动角度差距越大，则受测者越需要用更长时间去完成心理的转动，以便决定这两个物体是否是相同物体（图3-6a、b），还是不同物体（图3-5c）。

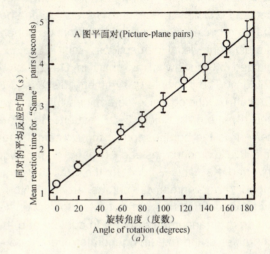

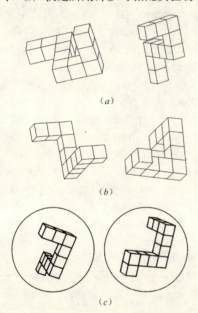

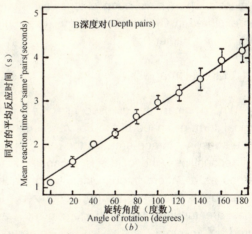

图3-6 平均反应时间与不同角度的函数
（a）相同的一对在图平面上转动不同角度；（b）相同的一对在深度面上转动不同角度
Figure 3-6 Average reaction time as a function of the angular difference
（a）The same pair differing by a rotation in the picture plane；（b）The same pair differing by a rotation in the depth. (Reprinted from "Mental rotation of three-dimensional objects", by R. W. Shepard and J. Metzler, *Science*, 171, p. 702, 1971, with permission from American Association for the Advancement of Science)

图3-5 测试心智转动的测试物
Figure 3-5 Stimuli used to test mental rotation (Reprinted from "Mental rotation of three-dimensional objects", by R. W. Shepard and J. Metzler, *Science*, 171, p. 702, 1971, with permission from American Association for the Advancement of Science) Models were regenerated by 3D modeler.

the context of subsequent words can provide more information for comprehending the sentence.

Context also influences visual perception. For instance, in Figure 3-2, we perceive five columns of circles rather than five rows, because the circles in a column are closer together than the circles in rows. Various contexts generate different visual perceptions, which are explained by the law of proximity in Gestalt psychology. The principle of symmetry discussed in Gestalt psychology also determines the way we see a pattern. In Figure 3-2, pattern B is perceived as the composition of three vertical lines and five diamonds instead of many normal and inverted Ks. This is due to the fact that diamonds are symmetrical but Ks are not. Non-symmetrical patterns are not easy to recognize. Different arrangements shown in context and the character of the patterns generate different perceptions, which have been extensively studied and explored in Gestalt psychology (see introductions in Chapter 2).

3.4 Mental image

When human beings see an object, the original form or shape of the object is imprinted in the sensory storage area. At the stage of pattern recognition, the object is recognized and learned as a pattern through identifying embedded features. After patterns are recognized, the codes are further interpreted, organized, and stored in memory (Kosslyn & Pomerantz, 1977). It would be interesting to determine how the recognized information is stored, and the formats of the codes stored in memory. This all relates to the study of mental images and visual images.

A **mental image**, as proposed by scholars, is a spatial representation analogous to the experience of seeing an object during visual perception. The mental image has size, dimension, color, and a certain level of texture, which represents the object in reality. After the object is seen, an image picture is established, which can be seen in the mind's eye as if visualizing the real object. This phenomenon suggests the notion of mental image scanning, which means that the mental image appearing in the mind's eye is analogous to the image perceived in one's vision. The analogy between mental pictures and visual images suggests the time it takes to scan from one object to the other object is the function of the real distance between the two (Kosslyn, Ball & Reiser, 1978). In their experiments, Kosslyn and his colleagues used a fictional island map that showed a tree, rock, hub, well, lake, sand area, and grass in the island (see Figure 3-3). Subjects were asked to picture the map in their mind and an object was named aloud for the subjects to focus on. Four seconds later, a second object was named and subjects were required to scan the map to decide whether the second object was on the map—by imagining a black speck moving from the first object to the second object within the map. When subjects reached and found the second objects, they pushed an "arrived" button to stop a clock used for measuring the **reaction time**. This showed the farther apart the two objects are, the longer the reaction time. In other words, as the distance between real objects increases, the scan time increases as well (see Figure 3-4) If the spatial relations between objects are non-linear, then the mental scan is non-linear.

Mental operations can also be three dimensional. For instance, Shepard and Metzler (1971) used two-dimensional drawings of three-dimensional objects to test human perception. In their stimuli (Figure 3-5), they showed pairs of objects in different orientations. Subjects had

在其他与心智影像运作有关的研究中，库柏和谢巴德（Cooper & Shepard, 1973）用正常和反置镜像的字母"R"图形（"R"图形是所用六个字母测验刺激物之一）来探讨心智的旋转功能。在他们的**实验刺激物**中，一些"R"是由正垂直被按顺时针方向旋转到六个不同的角度。图3-7即是所用六种字母之

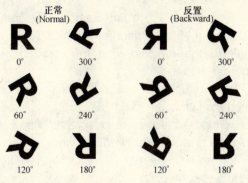

图3-7 字母转动受测物

Figure 3-7 Stimuli used by on letter rotation （Reprinted from "*Chronometric studies of the rotation of mental images*", by L. A. Cooper and R. N. Shepard, p. 95, 1973, with permission from Elsevier）

一的正常图（左两行）和反置图（右两行），六个图形显示所用刺激物的六种不同受测方位。实验中，受测者会看到两个连续的刺激物。第一个可能是在正常方位的字母"R"，第二个可能是正常"R"或倒置镜像"R"由水平转向某个角度。受测者必须决定第二个受测物是否能与第一个受测物相配。他们发现，受测者必须先在心中把物体转回到正直，然后再判断作比对，决定物体是正常或反置镜像"R"。花在正常图上的反应时间一般都持续地快于反置图。而且，回转角度越大，决定的时间也越长。最长的时间是在180°，过180°旋转，反应时间开始缩短。到360°时受测者的反应时间和0°相似（图3-8）。

这些视觉辨认方面的研究都证实了心智中三度空间回转物体的认知能力。然而，当物体被观看，心智影像被创立，而且被储存于记忆之后，这些影像即有些不同。因为心智影像并非心智图片，也不完全准确，更会被曲解或被分解成有意义的单元部分。这与记忆中心智影像的内在表示呈现法有关。亦即记忆中的影像储存是有某种特别的编码存在，但人类记忆的机能需要先讨论之后才能解释影像的内在呈现法。

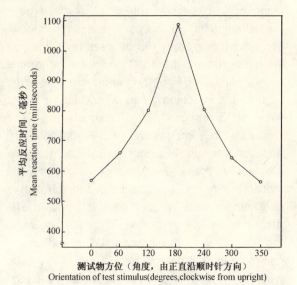

图3-8 评断字母垂直面转动角度对应函数的平均反应时间

Figure 3-8 Mean reaction time for judging the normality of a letter as a function of orientation angle（Reprinted from *Chronometric studies of the rotation of mental images*, by L. A. Cooper and R. N. Shepard, p. 103, 1973, with permission from Elsevier）

3.5 记忆

艾金森和雪弗林（Atkinson & Shiffrin, 1971）提出一个流程图指出信息的流动由周遭环境，经由感觉收录器传到短期记忆中。这传送行里有几项控制过程决定了信息是否该储存于长期记忆里，或仅仅是对目前的信息作出反应，输出对策。在这广被认同的观念中，有三个主要储存信息的区域被提到，即"感觉收录器"、"短期记忆"和"长期记忆"。**感觉收录器**是暂时存放数据之处，储存在这里的信

to decide whether these displayed objects were identical except for their orientation. For example, the pair on Figure 3-5 (a) is the same pair which differs by an 80 degree rotation in the picture plan (XY plane or Z axis, see note 3). Figure 3-5(b) is the same pair which differs by an 80 degree rotation in depth plan (XZ plane or Y axis, see note 3). In the experimental stimulus, all the pairs differed in orientation from 0 degree to 180 degrees in 20-degree intervals. Among the 1600 computer generated pairs, half (800) of the pairs were able to be rotated for matching, the other half (800) were mirror images that did not match (see Figure 3-5c). Images shown in Figure 3-4 (a) and (b) are just two possible matching examples among many. Experimental results showed that for every equal increment in rotation, an equal increment in reaction time is needed. The greater the angle of rotation differences between the two objects, the longer it takes subjects to complete the mental rotations in order to decide whether two patterns were identical (see Figure 3-5a and b) or not identical (see Figure 3-5c).

Other mental operation studies, Cooper and Shepard (1973) used normal and backward mirror images of character "R", which is one of six characters used to study mental function of rotation. In their **experimental stimuli**, some "Rs" were rotated clockwise to six different degrees from the vertical. Figure 3-7 shows normal and backward versions of one of the six characters. In this Figure, there are six orientations appearing as test stimulus. Subjects were presented with two successive stimuli. The first could be the letter "R" in its normal orientation; the second was either a normal "R" or a backward one rotated to some degree from the horizontal. Subjects had to decide whether the second stimulus could match the first one.

They found that subjects had to mentally rotate the image back to upright before judging it normal or backward. The reaction times were consistently shorter for the normal stimuli than for their mirror images. Further, the judgment times increased with the increase in the degree of rotation. The longest reaction time was for 180 degrees of rotation. When greater than 180 degree rotation, reaction time began to decrease until it reached 360 degrees, which was very similar to the reaction time for the 0 degree (Figure 3-8).

These studies in visual recognition demonstrated the subjects' cognitive ability to perceive three dimensionally rotating objects. However, after objects were perceived, the images created, and stored in memory were somewhat different. Because, these mental images are not mental pictures, they are not precise. They can be distorted and segmented into meaningful parts, which relate to the internal representation of an image. The images stored in memory have special coding properties. It is important to discuss the functionalities of human memory before explaining the internal representation of an image.

3.5 Memory

Atkinson and Shiffrin (1971) proposed a detailed diagram suggesting that information flows from the environment through sensory registers to short-term memory, where it is handled by several control processes that either store it in long-term memory, or simply react to it and provide some response output. In this well received concept, three important areas of holding information were discussed: sensory registers, short-term memory (STM), and long-term memory (LTM). The **sensory register** is the place that holds information that is temporary and is rapidly decaying. This notion was

息很快即会衰退。这观念被史朴令（1960）证明出。他的实验是给受测者经由**塔驰斯投镜**（见附记4）放出一排英文字母，时间是50毫秒，受测者看了影像后，即被要求把能记住的展出字母回报给测者。结果显示，展出的字母如果少于五个，则回报的准确度高。如展出字母数增加或受测者等待回报的时间期越久，则表现率就越低。这现象明示记得住的物体很快速的即衰散，尤其是在一秒之后几乎全忘记。这也说明了短促且有限的视觉储存器之存在。这区中记忆只能短暂但有效地保留一些看到的视觉信息。

菲利普斯（Phillips, 1974）也经由实验证明这感觉收录器的现象。在他的实验中，受测者要看一些方形矩阵，矩阵中的单元组集被任意填充形成某种形态。这些矩阵可能是五乘五或八乘八的矩阵组合。在实验过程里，第一个矩阵被展出一秒钟，然后停顿一下（停顿时间为20、60、100、300或600毫秒不等）再显示第二个，这第二个展示可能是相同的，也可能是相似的形态。受测者必须最后决定这两个展示的形态是同或异。结果显示停顿时间越长，准确度越低。这说明建议受测者是用感觉缓冲储存器来收录信息以备配对。在此实验中最有趣的是如果在两个展示中间的停顿期低于300毫秒，则受测者使用缓冲的结果之准确度越高。这表明了感觉收录持久的时间大约是1/4秒（250毫秒）。如此，感觉收录区具备收存信息以备工作之需的认知能力。当信息被注意到，短期记忆即开始起作用。

当人需要学习、作决定或解决问题时，**短程记忆**即能将环境里的信息和长期记忆中的信息合并处理。因此，短期记忆也被称为工作记忆。有两个因素限制了短期记忆的机能，即"记忆遗忘"和有限的"记忆广度"容量。记忆遗忘是指短期记忆中如无"复诵"，信息即失。如果信息不被复诵，则在短期记忆中立刻即被遗忘。皮特森（1959）要求受测者听并记住三个子音和一个数字，然后开始按序倒退计数，每次说出这数字减三的数，直到看到灯光亮，即回报这记住的三个子音。倒数三个数字的过程是防止受测者去复诵这三个要记的子音。灯亮的时刻是在受测者开始倒数之后依3、6、9、12、15及18秒的间距不等。例如，受测者听到字母C、H、J，跟着是506。则受测者要记住这三个字母，同时倒数说出503、500、497等，一直到灯亮即回报记住的C、H、J。实验结果证明三秒的延时，受测者可保证大约50%的正确率。18秒之后，受测者很少记得这三个字母。因此，正确的回报率是在回报时间18秒的期间里逐渐减低。这快速的遗忘率说明为了让短期记忆中的信息有效，人必须复诵。然而，这快速遗忘的程度比例最主要原因是由于受到干扰而生，并非完全是衰退的影响（Reitman, 1974）。

第二个限制短期记忆机能的因素是其广度或容量。记忆广度量即是一个人典型能回忆记出的最长次序量。有名的短期记忆广度之魔术数字"七加或减二"是由米勒（1956）提出的。米勒回顾许多关于记忆广度的发现并采取海斯（1952）之实验结果，即记忆英文单词的记忆广度量是五个单词，记二元数字（例如：001011101）的记忆广度量是九个项目。因此，米勒依此推断出短期记忆中一般记字母和数字的能量，是落在九和五中间的七，或者说是七加二或七减二。这也反映在世界许多国家的日常电话号码长度是七位数字。记忆广度量是七也可由下例测出。试着读图3-9中的四列字母，每次读一列。读完一列后，闭上眼睛以正确次序回记这读过

KJNRBLC

WPUYMXQSH

ZMQNPJK

OTGFREVDA

图3-9　能回忆到的最长可能字母

Figure 3-9　Possible characters of longest sequence for recall

demonstrated first by Sperling (1960), who showed subjects an array of English letters for approximately 50 milliseconds in a **tachistoscope** (see note 4). The subjects were then asked to report all the letters they could remember from the display. The results showed that responses were highly accurate if the display contained fewer than five letters. If the number of letters increased or the subjects waited longer to report, the subjects' performance decayed. It appears that the memory for reporting display items decays very rapidly and is essentially gone by the end of one second. This also indicates the existence of an extremely short and limited visual sensory store, a memory that can effectively hold all the information perceived in the visual display only briefly.

Phillips (1974) also proved the sensory storage phenomenon, in which subjects viewed patterns made by randomly filled cells in a square matrix. The matrix was either 5 by 5 or 8 by 8 cells. The first pattern was presented for one second and was followed after variable intervals (20, 60, 100, 300, or 600 milliseconds) by either an identical pattern or a similar one. Subjects had to decide, after viewing both, whether the two patterns were the same or different. Results showed that as the inter-stimulus interval increased, the accuracy declined. This suggests that subjects used the sensory buffer to store information for visual matching. The most interesting thing found in the experiment is that the use of the sensory store resulted in more accurate performance when the inter-stimulus interval was less than 300 milliseconds. It suggested that the sensory store lasts only about a quarter of a second (250 milliseconds). As such, the sensory store (register) has the cognitive ability to hold information. After the information is attended to, STM starts to function.

Short-term memory can combine information from both the environment and LTM whenever humans want to learn, make decisions, or solve problems. STM also has been called working memory. Two factors, forgetting and short **memory span** limit its functionality. Forgetting relates to the loss of information in STM without rehearsal. If information in STM is not rehearsed, it is forgotten rapidly. Peterson (1959) required subjects to hear and remember three consonants and count backward in intervals of three until they saw a light, which was the point at which they were to recall the consonants. The process of counting backward in intervals of three was to prevent subjects from rehearsing. The light was shown at 3, 6, 9, 12, 15, and 18 seconds after the subjects started counting. For example, subjects might hear the letters CHJ followed by 506. Then subjects should remember the letters and begin verbally counting 503, 500, 497, and backward, until a light was shown. At this point, subjects were to report back the letters CHJ. Experimental results showed that subjects remembered the three characters approximately 50 percent after three second intervals. After eighteen second intervals, subjects seldom remembered the characters. Accurate recall decreased over the 18 second intervals. The rapid forgetting rate indicates that humans must rehearse information in order to keep it available in STM. However, the rapid rate of forgetting is caused primarily by interference rather than decay (Reitman, 1974).

The second factor that limits the functionality of STM is its span or capacity. Memory span is the longest sequence in the typical recall that a person can make. The famous "magic number of STM span of 7 plus or minus two" was proposed by Miller (1956). Miller reviewed a large number of findings on memory span and

的字母，大部分人都可能回记七个字母（第一及第三列），而非九个字母（第二及第四列）。

为更进一步地解释这些信息是如何储存于短期记忆中的，米勒（1956）提出了一个名词"**组集**"（也称组块）以解释记忆里的知识组构单元或储存单位。他指出记忆受到的限制，不是以刺激物中真正的单元数量（如字母或单词）而定，而是由具有意义的组集数量而定。人能记得的信息大约是七个组集，这些组集也可能是存在长期记忆中的单元。每当这些单元被记起或被催化，即立刻备用。信息情报也可被重新编码形成新的记忆组集。只要这些组集是有意义的，则易于记忆。所有的信息也容易被学到、储存，并克服短期记忆的有限容量之束缚。下列 12 个字母组群是个好例子。如果 CI—AFB—IMC—IAT—M，这串字母依序成组的念出给听者，并要求听者依序回报，则回报率不会比念出 CIA—FBI—MCI—ATM 的组集更佳。因为第二个字符串有四个组集，每一个组集都为有意义的字母群。这说明把组集重组成有意义的团块更能帮助记忆。

记忆组集包存所有学到的知识单元，并成团块地学到、转移、并存放于长期记忆中。信息由短期记忆转存到长期记忆就意味着学习到并修炼成的结果。长期记忆并没有能量限制，能容存无限信息。这些信息都是被转成永久编码而完成的。也因为是永久编码，在长期记忆中的记忆衰退率也比短期记忆快速的衰退率要慢得多。一项与学习和编码效果有关的因素是"**复诵**"现象。复诵是把认知到的信息重复演练，以便保存的行为。朗都斯（1971）测验了复诵次数和可回记项目的关系。他以 20 个名词单词依序展示给受测者，每次一个词只展五秒钟。每次展完之后，朗都斯给五秒钟间隔要求受测者复诵所有看到的字。当所有字展完之后，受测者要不限次序回报所有记得的字。结果显示在名词单子开头的字都比其他字复诵次数要多，因此这些字已经转到长期记忆中，也有更高比率可由长期记忆中回

收。而在单子后面的字也因在受测者回报时仍然存在于短期记忆中，也易于立刻由短期记忆中抽取回报（图 3-10）。因此这实验中单子前后几个字的回收率都较高。图 3-10 中黑圈代表回记的可能性，白色方块是每个单词的平均复诵次数。

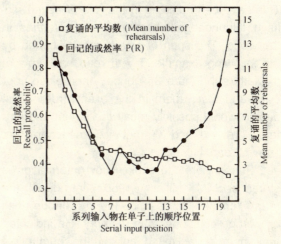

图 3-10 复诵次数与回记率的相互关系

Figure 3-10　Relations between the number of rehearsals and the probability of recall（Reprinted from "Analysis of Rehearsal Processes in Free Recall", by D. Rundus, *Journal of Experimental Psychology*, 89, p. 63-77, 1971, with permission from American Psychological Association）

有些理论指出信息存在于长期记忆中不会遗失，只是人们失去能力把它收回。这一现象涉及记忆的表示呈现和记忆内容是如何被保存的观念。最早以实验法探讨记忆是赫曼·艾宾浩斯（1850~1909）。他的理论说明心智是一个在单元之间有**联想**的网络。如果两个事件同时发生，其中之一必和另一事件在心中由某些关系连接住。在关系连接成立之后，元素即相连，回忆也可经由这联想而促成。这就是早期艾宾浩斯发展出的人类记忆的观念（Mook, 2004），也被更进一步地演化成**网络理论**（Anderson & Bower, 1973）或**语意网络理论**（Collins & Quil-

applied the experimental results obtained by Hayes (1952) that memory span ranged from five English words to nine items of binary digits (i. e., 001011101). Miller determined that the general STM memory span for letters or numbers is "7±2" falls in the middle of the range. This has been reflected in the use of seven digits in telephone numbers in many countries in the world. The span of seven can be tested by the following example. Try to read each row of the letters in Figure 3-9 once, then close your eyes and recall them in the exact order. Most people would possibly recall the seven characters (rows 1 and 3) but not nine characters (rows 2 and 4).

Furthermore, to explain how information is stored in STM, Miller (1956) introduced the term "chunk" to describe the knowledge units or storage components in memory. He indicated that memory is limited not by the number of physical units (letters or words) in the stimulus but by the number of meaningful chunks. Human can remember possibly seven chunks, which could be stored in long-term memory. Whenever chunks are activated or remembered, they are immediately available. Information can also be recoded to form new memory chunks. As long as these chunks are meaningful, they are easy to memorize allowing more information to be learned and stored, thus increasing the limited capacity of STM. A good example is the following string of twelve letters. If the string of letters, CI-AFB-IMC-IAT-M is sequentially read to a reader, and the reader is required to recall the string of letters in the same order, the recall rate will not be better than the group CIA-FBI-MCI-ATM because the second string has four chunks, each containing three meaningful letters. This demonstrates that the reorganization of chunks into meaningful groups could improve memorization.

Memory chunks consist of knowledge units that have been learned, transferred, and stored by groups in long-term memory. The transferring of information from STM to LTM signifies the results of learning. LTM does not have a capacity limitation. It has unlimited capacity for storing information which is permanently encoded. Because of the permanent encoding, the rate of decay in LTM is slow compared to the rapid rate of decay in STM. One of the important factors relating to learning and the effects of encoding is the phenomenon of **rehearsal**. Rehearsal is the action of repeating the perceived message for retention. Rundus (1971) tested the number of rehearsals related to the probability of recalling terms. He showed subjects a list of 20 nouns. The nouns were presented one at a time with 5 second intervals. Rundus required subjects to repeat aloud words on the list during the 5 second intervals. Afterwards, subjects were to recall the words in any order. Results showed that words at the beginning of the list were rehearsed more often than the other words; therefore, they were transferred to LTM and have a higher probability of being retrieved from LTM. Whereas words at the end of the list were still in STM when the subjects began the recall and were easier to retrieve from STM (Figure 3-10). Therefore, words in the front and the end of the list had higher retrieval rates. The black circles in Figure 3-10 represent probability of recall and the white squares are average number of rehearsals per word.

Some theories indicate that information in LTM is never lost, but humans might lose the ability to retrieve it. This phenomenon relates to the concepts of representation of memory and how memory contents are stored. The first to explore memory through experiment was Hermann Ebbinghaus (1850~1909). He theorized that the mind is a network of **associations**

lian, 1969; Collins & Loftus, 1975; Anderson, 1983)。这理论的观念是说成团的信息是组合在一体系中的不同层次,并以联想连接而成。就象征性而言,组集是节点,由环链连接而显示出组集间的关系。联想是指某一事件"甲"和另一事件"乙"之间有某些联系。因此在思考"甲"时,就会回想起"乙"。联想的形成有可能是在学习事件"乙"的过程中经过调整原有记忆"甲"团块而形成联系。因此当在回记团块"甲"时,经由激发在记忆网络中的联想引起蔓延联系,"乙"即被刺激回收起来。

根据网络理论(Anderson, 1980),记忆是一片网络由节点代表知识"组集"而形成。节点之间即以联想联结而成。当组集被回忆起,它们即呈活跃状也立即可被接触应用。所有节点被演变成活跃状态的过程称为触发刺激。要回忆在长期记忆中的信息,知识必须要先呈活跃可用状态。当信息已呈活跃状,在记忆中的该单元即进入活跃期,也代表短期记忆工作状态的形成,以便进行信息操作(这也是为何短期记忆被称为工作记忆)。在长期记忆中,回忆信息的过程是经由触动激化,经由连接扩展蔓延到长期记忆中所需要的记忆部位。如果联想有极强的联系,则相连的节点即易被回忆,否则即会失落。节点联结的强度可由"复诵"达到。这解释了在长期记忆中,知识和信息是如何被储存和回忆的,也说明了联系在记忆网络中所扮演的重要角色。

3.6 记忆增进术

通常,记忆会遗失,人也会忘记所记的信息。但即使是难于区分究竟失去信息是归因于信息没有被存好,还是归因于信息无能力被回收,也已经有方法(Norman, 1976)可以改进记忆和增进回忆。例如,在信息回收方面,恰当的线索可以协助寻找进入记忆中的信息。比方外在的记忆助理,如心理小记或心智影像等都会对回忆有帮助。其他可能的线索则包括活动性事件,以及和要回记事件相关或相似的事件等。这解释了如把关系连接给刺激起来,即会把与记忆组集的连接点亮。外在的暗示线索事实上提供了记忆组集新的连接,也更能把组集活络复苏。

其他增进记忆的方法也已经发展出。最明显的方法是在信息要被收入记忆中的时刻,创造出明确而且易记的回记线索,比方生动的影像即是一例。其他方法则包括使要记忆的材料之组织方法变得更有意义或增加复诵的时间,以及减少干扰等。一项有趣的研究证明了短的学习分隔时期会减少干扰。比方学生在睡前准备考试的成绩要比保持一段清醒时间再考的成绩好。这归因于在准备完考试之后入睡的学生会比保持清醒的学生收到较少反溯性干扰的缘故(Jenkins & Dallenbach, 2004)。但最有效的记忆增进方法则是使用生动的影像帮助记忆和回忆。

总之,长期记忆是把相关信息经由联想而组成团块。团块与团块结合又组成更大的网络。在这复杂的网络中要寻找单元,非得借力于其他媒介。回忆线索即是用来亲近这些网络的方法之一。至于遗忘会发生则是源于以下的原因:(1)信息衰退或误置而导致编码失败;(2)脑部损害或干扰而造成储存失败;(3)由于忘记回记线索或因记忆组集之间的弱连接而造成回收失败。

本章解释了与人类智慧密切相关的基本认知构件。其中,影像对设计领域造成了极大影响。因为设计者在做设计时,即会用到许多心智影像。但心智影像是抽象的,也不一定必然正确地反映出真实性。所以,心智影像和观念是如何储存于长期记忆中的呢?这就形成一个有趣的问题,这个问题与心智表示呈现以及掌握知识的过程有极大关系。这也是下章讨论的重点。建筑设计元素中,包括形态、颜色、质感等都是值得做科学化研究的领域,也更值得进一步了解人脑如何正面或负面地对这些信息作出应对。好的认识有益于帮助设计者做出更好的建筑及空间设计。

among elements. If two events occur at the same time, the one must be associated with the other one in the mind. The associations between events or elements are formed by learning. After an association is established, elements are linked which could be recalled through the association. This concept of human memory was originally developed by Ebbinghaus (Mook, 2004), which was further developed into **network theory** (Anderson & Bower, 1973) or **semantic network theory** (Collins & Quillian, 1969; Collins & Loftus, 1975; Anderson, 1983). The concept of this theory is that chunks of information are organized into levels of an hierarchy connected by associations. Symbolically, chunks are nodes connected by links to show relationships among chunks. Association means that event A has a certain connection with event B. Thus, while event A is under consideration, event B would be recalled. The formation of an association is possibly done by adjusting the existing memory cluster of event A while event B is in the process of being learned to build up the connections. When the memory cluster of event A is under recall, association spreads in the memory network to activate event B.

According to network theory (Anderson, 1980, p. 168), memory is a network of nodes representing knowledge chunks. Linking of nodes develops by association. When chunks are recalled, they are in active state that make them immediately accessible. The process of creating the active state for nodes is called activation. To retrieve information in LTM, knowledge must be activated first. When knowledge is activated, the units of that memory are in a special active state, which represents active memory of STM for information processing. The process of retrieving information from LTM is achieved through the spread of activation to the desired portion of LTM. If the association is a strong one, connected nodes can easily be recalled, otherwise they are lost. The strength of connection is achieved by rehearsal. This explains how knowledge and information in LTM is stored and recalled, it also illustrates the important role that association plays in memory network.

3.6 Mnemonics

Often times, memory is lost and humans forget information. Even though it is difficult to distinguish whether the loss of information is due to the information not being stored or due to the information being unable to be retrieved, there are methods to improve memory and increase recall (Norman, 1976). For information retrieval, appropriate cues can help gain access to the information in memory. For instance, external memory aides of mental notes or mental images can help. Other possible cues include activities or similar tasks relating to the to-be-recalled items. This explains the phenomenon that activation of associations light up the links to the retrieved memory chunks. External cues actually provide new links to the chunks and probably make the chunks alive.

Methods also have been developed for improving memory. A significant method is to create distinctive and easily recalled retrieval cues—such as vivid imagery—at the time when the information is encoded. Other methods include increasing the meaningful organization of the material, increasing rehearsal time, and reducing interference. An interesting study demonstrated that short spacing of study periods reduces interference. Students who studied for a test immediately before going to bed did better on the test than those students who studied and remained awake before the test. This is based on the theory that students who slept after

附记：

1. 在脑神经科学中，大脑被认为是持续地在进行信息处理，看到的景和听到的音都会被视觉器官及听觉器官接收、记录并合成于大脑的不同部分，然后在大脑的其他区域凝聚成一体创出外在世界的一个整体影像（Singer，2007）。
2. 全体宏观：此词来自于希腊文。意指关切全体而非分开个体去分析部分构件。也就是把看得见的部分，加上看不见却存在的东西整合作整体思考或处理。亚里士多德在《形而上学》一书中把一般思考原则综合说明成"整体多于部分的总和"。这用法与完形心理学的理论相似。见本书第1章第3节。
3. 图3-5（a）左边物体是由"平面图视场"沿X、Y和Z三个坐标轴线（X、Y、Z轴分别是楼地面上的长度轴、宽度轴和高度轴）转动-60°、-11°及95°，右边图形则转动-60°、-11°和15°。图3-5（b）中的左图是在"南向正立面视场"依Y轴转动（64°、-15°、25°），右图转动（64°、-15°、-55°）。图3-5（a）及图3-5（b）两对物体各自不同之处是3-5（a）右图是在高度轴线（即Z坐标轴）上顺时针转动80°，而3-5（b）则是右图在纵轴线（即Y坐标轴）上顺时针转动80°。这两图是由三维造模系统仿造谢巴德和麦兹勒的实验刺激物所用的图例。
4. 塔驰斯投镜或称T-镜，是一种可在相当短而准确的时间内将视觉刺激物暴露，瞬时闪发，并投影到眼前的实验装置。详情可见网页 http://www.sykronix.com/researching/tscope.htm

studying would get less retroactive interference than would the students who remained awake (Jenkins & Dallenbach, 1924). However, the best mnemonics method would be the use of vivid imagery for memory and recalling.

In summary, long term memory is organized into clusters of related information connected by association, and the clusters are organized into larger networks. In order to search for items through the complex clusters, certain assistance is needed. Retrieval cues could be a method used to access to these networks. Forgetting can occur whenever there is: (1) encoding failure due to decay or misplacement; (2) storage failure due to physical brain damage or interference; (3) retrieval failure due to the forgetting of the retrieval cue or weak linkage between memory clusters.

This chapter explained the fundamental cognitive components essential to human intelligence. Among them, image contributes significantly in the field of design, because designers utilize many of mental images during their design processes. But mental images are abstract, not necessarily and precisely reflecting reality. Then, how are mental image and concepts stored in LTM? This interesting question relates to the representation of and process of handling knowledge, which are the topics to be discussed in the following chapter. In general, design elements such as shape, color, and texture are research areas that still need scientific understanding of how and why the human brain works with these elements. A better understanding can help designers in designing and building spaces.

Note:

1. In the field of neuroscience, the brain is recognized as constantly processing information; sights and sounds are received by visual and auditory sensors, recorded and synthesized in different parts of the brain, then fused together in other areas to create a cohesive picture of the outside world (Singer, 2007).
2. Holistic: A Greek word concerned with the whole rather than analysis or separation into parts. It means to think or handle the tangible components together with intangible but existing parts inclusively. The general principle was summarized by Aristotle in the Metaphysics: "The whole is more than the sum of its parts." Its meaning is similar to the concepts developed by Gestalt psychology, see, Chapter 1-3.
3. The left object on Figure 3-5 (a) has been rotated -60, -11, and 95 degrees along X, Y, and Z axes (X, Y, and Z axes are the length, width, and height axes on the floor plan); whereas the object on the right is rotated -60, -11, and 15 degrees respectively from the top plan view point. The pair shown in Figure 3-5 (b) were rotated (64, -15, 25 degrees) the left object and (64, -15, -55 degrees) the right object along the Y axis from the front election view point (south side view). Differences in the pairs: Figure 3-5 (a) pair has rotation difference in the height axis (Z axis) and the Figure 3-5 (b) pair has 80 degree difference in the width axis (Y axis). These views were generated from 3D solid modeler to match the two figures used in the experimental stimulus by Shepard and Metzler.
4. A tachistoscope or t-scope, is a device that can exposes, flash, and project visual stimuli before our eyes for very small and precise amounts of time for experiments. More information can be found at: http://www.sykronix.com/researching/tscope.htm.

第4章 心智影像与认知运作过程

　　思考是人获得宇宙知识的起源，也基于拥有这些知识，人才会对这宇宙有认知。认知是一种思考形式，也依凭此形式，知识在脑中被演绎，也被处理，以便满足某些企图和目的。因此，思考是一种错综复杂的心智活动，涉及不同的活动运作，包括影像、语言、抽象符号，以及动作等。其中，影像有别于语言，是独特的知识，它有特殊的**表征**（或称内在呈现）形态。表征的意思是指以某一物替代另一物，同时也是对发生在真实中真物的一种代表方式（Echenique, 1972; Hesses, 1966）。由此而言，影像不同于储存在记忆中的语意信息，本身就有一种独特的呈现款式存在。尤其心智影像会是另类思考形式，有别于也区别于其他的认知过程（Kosslyn, 1983）。本章将探讨和设计有关之心智影像的表征形态（呈现表示）、心智运作和记忆储存法。一系列心理实验详细解说心智运作，以及相关的研究观念与方法也将在本章中详述。

4.1 与设计有关的心智影像

　　正如第2章中所解释的，设计是一系列为满足某些目的而执行的活动。在这过程中，设计者被要求在基本逻辑推理上具有某种程度的信心，也要具备解决设计问题的一些策略。建筑设计无论与其他专业有多大的差异，最基本的专业都是依赖一套和建筑体块、机能及文化含义相关的知识。这套知识也要求一些图形和体块的几何数据，加上对颜色及材质的特殊智能，这些知识信息成为视觉影像和某些抽象内涵合并而成的一体。所成的**心智图像**也是某些不真实或不存在物体的一些**心智图片**，称为心像。它也是心智把一个实体的造型在心里创造、重建或模仿之后所产生的影像。

　　对设计师而言，使用心智影像来帮助思考，已不是一个新课题。设计师通常在心智里操作基本几何形体，而后将操作结果以几何图形或实体模型的形式将设计外显重现。换言之，素描草图通常会先在心里发展出来，然后浮现在绘图桌上。一位设计师无论在何时面临一个设计，他或她的第一个反应是在脑海里想像一个潜在的解决方案，并将其具象化（Porter, 1979），这就是心像。许多有名的科学家、作家、发明家和艺术家都同意心智影像会帮助也会增进他们的创造思考能力（Shepard, 1978）。

　　一些研究也指出心智影像在设计中所扮演的角色。达克（Darke）指出在设计期间，设计师会在他们心中发展出一个关于客户（或乙方）的综合影像以提供设计推测的数据来源。这影像由一个视觉观念发展而出，而这视觉观念则是由几个简单的设计目标诱导而成（Darke, 1979）。其他的研究也明示一个巨大影像库在记忆中的存在，这影像库由体验

Chapter 4 Mental image and cognitive processing

Thinking is the basis for knowledge of the world and for reacting to the world based on that knowledge. Cognition is the form of thinking in which we internally interpret knowledge and process it to satisfy certain purposes. Thus, thinking is a complex activity that involves operations of knowledge of images, language, abstract symbols, and actions. Image, which differs from verbal, is very unique, with special internal representation. **Representation** means to have something standing in for something else and is the means for representing the things that happened in reality (Echenique, 1972; Hesses, 1966). Image, in this regard, might have a special and unique format of representation different from verbal information stored in memory. Particularly, mental imagery could be a mode of thought separate from other cognitive processes (Kosslyn, 1983). This chapter will explore representation, mental operation, and memory storage of mental imagery particularly in relation to design. The results of a psychological experiment that explains the cognitive operations and research concepts and methodologies will also be covered in this chapter.

4.1 Mental imagery in design

As explained in Chapter Two, design is a series of actions executed to serve certain purposes. In the process, designers are required to possess competence in basic logical reasoning and in strategies for resolving design problems. The essential expertise in architectural design, however distinct from many other tasks, relies on a body of knowledge about architectural objects, functions, and cultural meanings. Also required is knowledge of geometric information on shapes and forms, and special knowledge of colors and textures, which combine certain visual images with certain abstractions. These visual images are **mental pictures** of something not real or present and are called **mental images**, which are a mental creation, reproduction, or imitation of the form of a physical object.

The use of mental images to help thinking is not new to designers. Designers often mentally manipulate the spatial relationships among fundamental geometric shapes and then present the product of this manipulation in graphic or model representations. In other words, sketches are tried out in the minds'eye before they e-

到的地点及事件之影像组成，而且因人而异（Downing, 1992）。这些影像也提供了联系，把设计课题和经验影像连成一起。所有这些研究都说明了视觉图像不仅是设计知识的一部分，并也用于解决设计问题。但这些心智影像是如何组织并存放于记忆里的呢？这些问题在下列章节中仔细探讨。

4.2 记忆中的心智影像

画家画山的方法是到山里去看山，在看山读山的过程中，画家会把山抽象化，之后，一个山的形态或原形就开始发展出来并记在脑海中。这发展出的形态不只具有心智影像的造型，并且也是画家对山的观念和想法。这种脑海中的形态也曾被称为**基模**或知识模式（Gombrich, 1964），也就是一种知识的表征形态。这基模的成形，使得影像之形态可被使用，而且画家也能画出他自己的山，而非任一特定的山。但在画家的脑海中，有成千的形态可用，并形成他个人的绘画数据库。

要研究知识的表征形态（或呈现表示）或基模的成形，从信息处理学的角度而言，必须要以心智处理数据的过程为核心，由感知被注意到的知觉刺激诱因起始。一般而言，刺激诱因中所存在的特征和对输入心智中的感觉之可能响应或对应，都被称为信息情报。把这些信息情报转成一种呈现的形态储存于记忆中的过程，就是编码。被编了码的信息会暂时居留在短期记忆里，经过复诵之后，转到长期记忆中作永久储存。人们会记忆到信息是因为在物体和数据代码之间建立联想而形成。所谓**数据代码**和刺激诱因的特征有关，也与记忆中被激发的可能潜在对应有关。当一数据代码被唤起后，相连到的相对物件就会被回忆起（Levy, 1971）。因而，与记忆有关的三个信息处理过程通常会被描述为记忆编码、记忆储存、记忆回记。

（1）记忆编码。记忆编码中的不少理论，都集中在何种代码与何种知觉刺激有关的问题上（Lea, 1975; Kosslyn, Ball & Reiser, 1978）。语言信息在脑海里以与语法相似的代码编成，这不同于因视觉刺激而生的视觉码（Levy, 1971）。视觉码编存着物体的大小、方位、颜色和相关的其他属性，同时也系统地组织成单元（Neisser, 1967; Reed, 1974）。布鲁克斯（1968）曾用实验证实在同时执行视觉和语言课题时，视觉码式和语言码式是绝然不同的。也有不少研究显示在长期记忆中，视觉材料的超强储存能力强过语言材料的储存能力（Shepard, 1967; Nickerson, 1968; Standing, 1973; Rowe & Rogers, 1975）。佩维奥（1971, 1986）就发展出了双码理论。双码理论是说，当人们经历到图片或文字时，即同时发展出视觉图片代码和语言文字代码，但视觉代码是特别为图片形成，而语言代码是为文字形成。这建议了人类记忆可能在感知的过程里具备着掌握双码的能力。视觉影像同时也有益于执行逻辑事件，如解决线性三段演绎的推理问题（Carramazza等, 1976）。一个左半脑受伤的日本病人失去了读平假名（即语音文字）的能力，但仍可读片假名（即象形文字）（Sasanuma, 1974）。这些发现都指出语言和视觉在记忆中的辨认系统，不但在机能上不同，而且在脑中存在地区的部位也不同，而且视觉表征在解决某些问题的课题里还是首要的解决工具。

（2）记忆储存。人类记忆中有三个储存机能：感觉储存区、短期记忆和长期记忆（见第3章中第3~5节）。长期记忆里组集的组织与人类如何保留与宇宙相关信息的态度有密切关系。在20世纪80年代发展出的**语意网络理论**，即完整地解说了信息在脑海中的节点环链相连之结构（见第3章中第3~5节记忆论）。语意网络理论代表一般性的内在智慧。但对于影像表征而言，一些研究也指出一个心智影像有可能是抽象概要的，并有两度建构的异质同构现象存在，它不但包括了空间情报，也以更抽象的形保存了物体的空间关系（Norman & Rumelhart,

merge onto the drawing board. Whenever a designer is confronted by a design project, his or her initial reaction would involve a visualization of potential solutions (Porter, 1979), that is, a mental image. Many famous scientists, writers, inventors, and artists also claim that visual imagery enhances their creative thinking (Shepard, 1978).

A number of studies have suggested the role that image plays in a design. Darke indicated that designers have a generalized image about users in mind during their design period. The image is generated by a visual concept derived from a few simple objectives to provide a source for conjectures in design (Darke, 1979). Other studies had shown the existence of a large set of image data bank in memory, consisting of places and events from experience unique to each individual designer (Downing, 1992). These images provide links that connect the design tasks and images of experiences. These studies suggest that visual imagery is a part of design knowledge used in solving design problems. But, how are these mental imageries organized and stored in memory? These questions are carefully explored in the following sections.

4.2 Imagery in memory

One way a painter learns to paint a mountain is to go out to the field to observe a mountain. The process of observation would abstract the form of the mountain, and then a pattern, or say a prototype, is developed and stored in memory. Such a pattern not only has the form of a mental image, but also is the painter's concept about the mountain. The term "**schema**" is used to describe the pattern stored in memory (Gombrich, 1964) which is a knowledge representation. The formation of the schema makes a pattern available, and the painter is able to draw his own mountain which may not be any particular mountain. However, in a painter's memory, there are thousands of patterns of objects that form a data bank.

The study of knowledge representation or formation of a schema, from the information processing point of view, must center around the mental processing of information, starting with the perception of attended sensory stimulus. Features of a stimulus and a response of sensory input are referred to as information. The process of translating this information into representational form stored in memory is encoding. The encoded information temporarily resides in short-term memory and after rehearsal it is transferred to long-term memory for permanent storage. People memorize by establishing associations between objects and codes. **Codes** refer to the features of the stimulus and potential response activated from memory. When a code is activated, objects corresponding to it are recalled (Levy, 1971). The three memory-related information processing stages are commonly described as: encoding, storage, and retrieval.

(1) **Memory Encoding**. Theories in memory encoding revolve around how a type of coding relates to a type of sensory stimuli (Lea, 1975; Kosslyn, Ball & Reiser, 1978). Verbal information is encoded in memory by speech-like coding that differs from the visual coding for visual stimuli (Levy, 1971). Visual codes encode information about size, orientation, color, and other properties of objects and are systematically organized into meaningful parts (Neisser, 1967; Reed, 1974). Brooks (1968) demonstrated that visual encoding is distinct from verbal encoding in performing both visual and verbal tasks. There also are a number of studies showing the superior capacity of long-term memory for holding visual material as opposed to corresponding verbal material (Shepard, 1967; Nickerson, 1968; Standing, 1973;

1975; Kosslyn & Pomerantz, 1977; Kosslyn, 1990)。虽然在目前并没有决定性的显示出人类长期记忆中的知识结构，可以节点环链相接的表征来正式代表，但许多用计算机仿真人类认知的研究都用这结构作为代表记忆的基本方法。

（3）记忆回记。由于短期记忆的有限能量限制，只有最当前被留心到的信息情报才能立刻被扣取（Ericsson & Simon, 1980）。因此，当长期记忆里的信息被刺激之后，即会转移到短期记忆中备用，也会因非活跃而失落。但因为信息情报是组集构造，因此内在知识的构造可由活跃存在于短期记忆中的组集（组块）来探讨。换言之，当记忆网中的任一情报组集被刺激后，即被回记起并转移到短期记忆中。探讨短期记忆中的组集即可提供线索说明知识的表征构造。这研究方法可由时间决定，因为在内在知识的表征里，回记数据所需的时间即是由长期记忆中回取组集的时间，也被称为是**反应时间**。如果被照顾到的组集数据需要输出外显，则其必得要通过一些通道（或**效应器**）以便由短期记忆中剥离。因为输出的通道有交通能量限制，因此组集中的数据必须以线性，一对一的方式从效应器中转移。因此，在长期记忆中的两个数据情报中间的停顿期即提供很好的线索来度量转移的耗费时间。如在外输两个情报间的耗费时间越长，即是跨越两大组集的象征。时间越短，即表示这些输出的数据是在同一组集里。这分辨数据单元的方法曾被却斯和司马（1973）用来研究西洋棋的认知现象。在他们的实验中，受测者先观看一局西洋棋的棋盘，然后被要求回忆棋盘并依序复原棋子的布局。他们即测量并记录连续回记棋子的时间，发现两个组集之间的界限大约是两秒钟。在他们其他的实验中，他们也发现长顿时期和错误大都发生在新组集的起始时刻（1974）。

4.3 记忆中影像呈现的双码论

心智影像是另一种形式的智能知识。设计师通常在设计案中都会引用影像做设计，也因此逐渐建立起一个特殊的个人设计知识体系。一个被公认的设计知识体系由专职教育过程中学到，也由专业训练中发展而来。设计大师都有一大组设计技艺清单储存于他们的记忆中，这也如同上了段数的棋王大师能记得住成千的棋谱，或音乐大师能记得住上万的音律一样。但无论如何，心智影像必有其特别的双码形态之表征存在于脑海里。这可由建筑设计师的建筑设计知识来解释此现象，因为建筑设计要求许多二维及三维影像的使用。

建筑设计知识是一专业的**领域知识**，其知识结构也和其他专业知识（如会计师）稍有不同。如由**语意网络理论**这方面来解读其记忆表征，则每一个知识组集（或节点）都有一文字符号来作识别卷标，这是文字码。一个组集（节点）可能是一个设计单位，代表一个实际的设计构件，并拥有该构件必须具备的属性值存在。一群组集（节点）会凝聚而形成一个概念，并构成一个巨型组集（节点）。在每一组集（节点）里，更有细部组集（节点）代表其单元组构。总体而言，整个组集（节点）的结构一方面如同一颗树的形态，但在某些组集间又以联想互连而形成一个巨大网络。在树结构里高层次组集（节点）的文字码卷标比低层次代码更为抽象。且在树叶部位的基层原点组集也比其他高层次的组集（节点）有更具体的意象，也因此比其他组集带有更特殊的视觉码。

图4-1列举了一个客厅设计观念的语意网络，这一客厅知识组集（节点）具有许多低层次组集（节点）。组集（节点）之间的关系由建筑机能（或功能）联系而成。在这背景脉络里，建筑机能是指一个建筑物的自然、位置、关联、结构、技术、可行性以及使用的设计要求（见附记1）。因此，组集（节点）建筑机能互连的概念涉及物体与物体之间自然、位置、连接、结构、技术和使用上的相互关联。也因为有建筑机能，所以物体的组集（节点）能连

Rowe & Rogers, 1975). Paivio (1971, 1986) developed the dual-code theory that an encounter with either a picture or a word may develop both a "visual" and a "verbal" code. But the visual codes are formed for pictures and verbal codes are formed for words. This suggests that human memory may be equipped to handle both codes developed during the process of perception. Visual imagery also is instrumental in performing logical tasks such as linear syllogism problems (Carramazza, et al., 1976). A Japanese patient with left-hemisphere damage lost the ability to read kana (phonetic characters) but was still able to read Kanji (ideographic characters) (Sasanuma, 1974). These findings suggest that verbal and visual recognition memory systems are functionally and topographically distinct, and that **visual representation** is used primarily as a tool in certain problem-solving tasks.

(2) **Memory Storage.** There are three storage functions in human memory: sensory storage, STM, and LTM (see Chapter 3). The organization of chunks in LTM relates to the manner in which humans retain information about the world, and the **semantic network theory** developed in the 1980's well explained the node-link structure (see Chapter 3, section 3~5 on memory). The semantic network theory represents internal knowledge in general. For image representations, some research suggests that an image may be an abstract schematic and that it has a second order isomorphism, containing spatial information and preserving spatial relations of objects in a more abstract form (Norman & Rumelhart, 1975; Kosslyn & Pomerantz, 1977; Kosslyn, 1990). It has not been conclusively shown that knowledge in human LTM can be represented formally by such node-link structures, but many computer simulations of cognition use such a structure as the fundamental means to represent memory.

(3) **Memory Retrieval.** Due to the limited capacity of our STM, only the most recently heeded information is accessible directly (Ericsson & Simon, 1980). Thus, when information in LTM is activated, it will be transferred to STM for use and may be lost when it becomes inactive. Because information is formed by chunks, the structure of internal knowledge representation can be explored by studying the activated chunks in STM. In other words, a chunk of information residing in a certain area in the memory network is activated during memory retrieval and loaded into STM. The study of chunks in STM could provide clues to understanding the structure of knowledge representation in LTM. The method is through measuring retrieval time. The time needed to retrieve information from internal representation is the time needed to retrieve chunks from LTM and is recognized as **reaction time**. If the loaded chunk of information has to be externally presented, it must pass through a type of **effector** channel before it is removed from STM. The output channel has limited capacity for transferring information, so information in chunks must pass through the channel in a linear, one-by-one output fashion. Therefore, the time of pause between two pieces of information retrieved from LTM provides a good measure for the transmitted latencies. The longer latencies of retrieval time between two pieces of information symbolize the boundaries between chunks, and the shorter latencies between pieces indicate these pieces are within the same chunk. This method of detecting units of information was used by Chase and Simon (1973) when studying the same perception in chess. In their experiments, subjects were asked to recall chess positions after viewing a chess game display. They measured

成一体，形成建筑知识表征。简言之，图4-1中每一组集（节点）都带有文字码和视觉码。高层次组集（节点）比低层次组集（节点）更抽象。

图4-1　一个客厅的影像知识表征图
Figure 4-1　An image knowledge representation for a living room

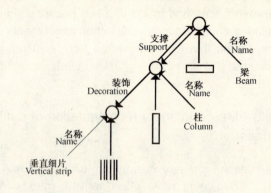

图4-2　柱、梁的文字码和视觉码结构
Figure 4-2　The structure of visual code and verbal code for a beam and a column

例如，柱和梁在建筑网络中算是初级节点，具有清楚的视觉码，如图4-2所示。图4-2节点之间是由建筑机能连成，这联系可由命题"A（X，Y）"定出。这命题中X代表行为者，Y代表被行为到的对象，A代表二者之间的关系。如此，柱梁之间的关系即可表达为"支撑（柱，梁）"或"坐置（梁，柱）"。在柱身上垂直的凹片装饰亦可定为"装饰（垂直细片，柱）"。所有这些支撑、坐置和装饰代表两对象的关系，也形成网络中底层节点间的联系。于是，节点间逐一地由关系互连而逐渐形成整个视觉网络。

进一步而言，储存在长期记忆中的知识，是假定基于对象精通而把该对象的相关信息组集化的产品。成名的设计大师，由于有更丰富的建筑机能、关联和产品规格经验，不但更能把信息情报依建筑"机能"储存，并且也因此集有更多的建筑信息。所以，专家设计师如图4-3和图4-4所示，应有更丰富的整套知识表征。图4-3是壁炉观念，图4-4则是更详细地以图例说明文字码和视觉码如何联结，形成一个客厅中的壁炉代表。不过图4-3和图4-4是在理论上代表一个平常的壁炉，并不定位为某一

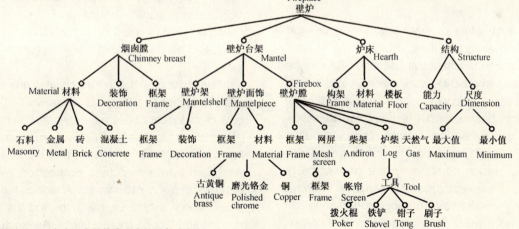

图4-3　一个在语意网络中丰富的壁炉表征
Figure 4-3　A rich set of semantic network representation of a fireplace

the time intervals between placements of successive pieces and found that two seconds signified a chunk boundary. In other experiments, they found that long pauses and errors usually occurred when new chunks were formed (1974).

4.3 Mental imagery representation of dual codes

Mental imagery is another form of knowledge. Designers have often applied images in their project exercises which gradually build up a specialized body of design knowledge. As recognized, a body of design knowledge is acquired through a process of professional education and developed through professional training. Design masters have a large set of design repertoire stored in their memory, which is the same as chess masters who have memorized large sets of chess game patterns, or music rhythms by grand musicians. However, the mental imagery might have a special format of dual-code representation in memory that makes it special. This could be explained better by architects' architectural design knowledge, because architectural design projects require use of many two dimensional and three dimensional images.

The structure of architectural **domain-specific knowledge** differs from other domain knowledge (for instance accounting). Approached from the **semantic network theory**, each knowledge chunk (node) in the knowledge representation has a verbal symbol for identification, which is the verbal code. A chunk (node) could be a design unit representing a physical design component that has some attributes of its own properties. A cluster of chunks (nodes) would have grouped together to form a concept, which is a mega chunk (node). Within each chunk (node), there are sub-nodes representing components of the chunk (node). The entire structure of the chunk (node) is a tree format on one hand but linked through chunks by association to form a huge network. Verbal symbols in higher-level chunks (nodes) in the tree structure are more abstract than are lower ones. Primitive chunks (nodes) occupying the leaves of the tree structure tend to have more concrete meanings than all other nodes and, thus, they possess more specific visual codes than others.

Figure 4-1 is an example that shows the semantic network of a living room concept, which is a chunk (node) of knowledge that has many lower level chunks (nodes). The relationships between chunks (nodes) are connected by architectural functionalities. In this context, architectural function means the nature, position, connection, structure, technology, feasibility, or usage required for designing an object (see note 1). The architectural functional relationships connecting chunks (nodes) are thus referred to as the associations among objects that are correlated by nature, position, connection, structure, technology, or usage. It is the architectural functionality that associates objects together in the knowledge representation. In short, every chunk (node) in Figure 4-1 has a verbal and visual code attached. Higher (nodes) nodes are more abstract than lower ones.

For instance, beam and column are counted as primitive nodes in the network, which have clear visual codes as shown in Figure 4-2. The relationships between nodes in Figure 4-2 are connected by architectural functionalities, which can be denoted by a proposition with the form of "A (X, Y)", where A stands for a relation, and X and Y are the actor and the object being acted upon, respectively. As such, the relationship between the beam and the column

位设计师的个人定义。因为不同的设计者有不同的心智影像,也有不同的知识表征。但在知识结构上的研究,则会提供数据,探测设计师如何组织他们的设计知识。

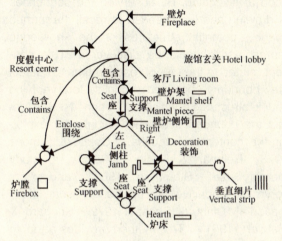

图 4-4 由文字码和视觉码连成的壁炉之知识表征
Figure 4-4 A knowledge structure of a fireplace connected by both visual & verbal codes

4.4 心智进行影像处理的实例

图 4-3 所示的壁炉知识表征可能提供线索说明影像是如何被记住的。即使这结构是假设的,但其表征呈现可更进一步地借着由设计者在心里产生壁炉心智影像,并将这影像经由外在展示法将其外显来证实其知识组集之内涵。最理想的外显出心智影像的方法是素描手绘,因为一般设计者都精于素描。如此,经观察设计师如何绘制该壁炉心智图的次序过程和观察回记这影像形成的时间,该影像的知识表征及可能结构都可大致探知一二。这一观念涉及使用"反应时间"去探测组集的界限来证实组集结构,曾被用来测定组集与组集间的界限以探求组集的内涵(Chase & Simon, 1973; Reitman & Rueler, 1980);也被用来研究记忆的大小(Sternberg, 1967);更被用来观察视觉信息情报中的认知过程

(Clark & Chase, 1972; Shepard & Metzler, 1971)。因此,这方法提供了一个研究工具,探讨设计师的设计知识结构。

反应时间:

探测反应时间的观念与有条理、按部就班依序收取信息的明细程序有关。这过程可分解成下列几个详细的步骤:

(1) 当一个人遇到一个刺激信息,要回记一个特殊对象时,一个符号即被编码到短期记忆中作参考依据。直到该对象被完全记出回收为止才消失。这**编码**动作需时 E。

(2) 被编了码的符号即代表是在长期记忆中要搜寻的目标。当事人将会在长期记忆中搜寻具有该对象符号的组集,触发、激动这组集以便将它转移到短期记忆中。这在长期记忆中**回记组**集的行动需时 A。

(3) 如果信息回记成功,它即被转到短期记忆里,以便补充短期记忆中的最新数据。这**转移**行动需时 B。

(4) 在信息装备到短期记忆中之后,即需解码转成绘图格式。负责译码的认知组构是心智扫描。即在心里扫描已回记的影像,以便决定该对象影像的可绘尺度大小。**扫描**译码需时 C。

(5) 在心智影像扫描完成,任一绘图行动(执行并)结束后,知觉系统即察看图上已绘出的图形特征以便决定下个记忆回收之对象。所需**察觉**时间是 D。这解码和认察是循环递归性一直等短期记忆中所有的信息用竭为止。只有当短期记忆中已无符码,下一个长期记忆中的符码回记才能开始,下个轮回才能启动。

这从想像一个图片到画出这个图片为止的整个执行过程,是绝对权威的操作程序。亦即过程中每一步骤必得在其前的步骤完全结束之后才能启动。"反应时间"是真正执行心智课题所需的时间,而且每一个步骤都要有一段确定时间来完成该课题。如

can be expressed as "Support (Column, Beam)" or "Seat (Beam, Column)". The vertical strip decoration on the shaft of a column is, "Decoration (Vertical strip, Column)". All these relations of support, seat, and decoration make up the links between lower level nodes. Thus, nodes are connected with one another by relationships and the whole network of image can be gradually constructed.

Further more, knowledge stored in LTM is presumably the products of chunking information based on familiarity. Famous expert designers, who are more experienced on architectural functions, relationships, and products' specifications, are also presumably more likely to organize information according to architectural "functions" and to have a larger saved amount of architectural information. Expert designers might have rich sets of knowledge representation as shown in Figures 4-3 and Figure 4-4. Figure 4-3 is a rich set of concepts on a fireplace, whereas Figure 4-4 is a more detailed diagram suggesting how visual and verbal codes of nodes are connected to represent a fireplace in a living room. Figures 4-3 and Figure 4-4, however, theoretically represent a fireplace, which should not be construed as a definition of any individual designer. Different designers have different mental images and different knowledge structures. Studies on knowledge structure could provide information on how designers organize their knowledge.

4.4 Example of mental processing of imageries

The knowledge representation of the fireplace shown in Figure 4-3 can provide clues to explain how images can be memorized. Although the structure is a hypothetical one, its representation could be further explored by having designers generate mental imageries and output their images through external display to verify knowledge chunk contents. The ideal means for outputting mental imageries would be through drawing, because designers are usually good at sketching. By observing the sequences of drawing and time needed to recall the formed image, potential structures of the knowledge representation could be discovered. This concept relates to the use of **reaction time** to detect the boundary between chunks for verifying chunk structures, which has been used to detect the boundaries between chunks for exploring the contents of chunks (Chase & Simon, 1973; Reitman & Rueler, 1980), to study memory size (Sternberg, 1967), and to observe the cognitive processing of visual information (Clark & Chase, 1972; Shepard & Metzler, 1971). This method provides a tool to study the structure of designer's design knowledge.

Reaction time:

The notion of detecting reaction times relates to the detailed sequential processes of retrieving information which could be categorized into the following processing stages.

(1) When a person encounters a stimulus for retrieving a particular object, a label of the stimulus will be encoded in STM for reference until it is successfully retrieved. The **encoding** action takes time E.

(2) The encoded label will be the search target in LTM. The person will search through the LTM to find the chunk that contains the right information, activates it for ready to transfer it into STM. The time needed for **retrieval** from LTM is A.

(3) Upon being successfully retrieved, information will be transferred to STM to update its contents. The **transferring** action takes time B.

(4) After the information is loaded into STM, it needs to be decoded into drawing for-

果一个课题需要几个心智操作,则其反应时间是所有这些运作的总和。依此而言,脑中操作心智影像的反应时间可划分为:概念间、组集间以及组集内三大类。

①概念间=A+B+C+D+E。一个概念可看成是一个巨大(许多)的组集。回记一个概念的反应时间可看成是概念与概念间的时间。比方说窗、门和壁炉是三种不同的建筑形态。每一种都各自有许多次级单元。这些次级单元的影像在知识结构中是群聚成巨大(许多)组集体各自代表窗、门和壁炉的概念。在自由回收过程中,要回收的**观念体**影像之名称必先得记住。这是由落实文字码,将其编在短期记忆中而告成,需时 E。要激起长期记忆中符合短期记忆里文字码的影像组集,需时 A。把它转到短期记忆,需时 B。存在于短期记忆中受激的影像是抽象形态,需时 C 去扫描,以便决定所绘尺寸。于是绘图动作开始。在绘图动作完毕之后,知觉系统认察在图形脉络里绘成的特征,以便决定下一个行动,这查证行为需时 D。如果下一行动是要回记另一概念体,则新的文字码即成受测物的信号,提示要复诵以便编码到短期记忆中。而后整个回记过程重新开始。因此,概念体与概念体间相隔的反应时间是由知觉回记情况(D)、编码于短期记忆(E)、唤醒长期记忆中组集(A)、转移组集到短期记忆(B)、译码(C)、到开始绘图那一刹那的全部动作所需时间,亦即 D、E、A、B 和 C 的总和(图 4-5)。

②组集间=A+B+C+D。在概念体回记之后,概念中的组集即被触动。在处理第一个组集时,该组集中的单元部分即逐一地被唤醒催化。其代表符号不需复诵即可被自动编码,于是信息立即可用。组集单元即可逐一地绘出。至于组集间的反应时间是在该组集的影像被画完之后,知觉系统必得察看结果,决定进行概念体中下个适当的组集(时间 D),并由长期记忆中回记(时间 A)转到短期记忆(时间 B),经心智扫描影像中的物体以决定绘出的大小(时间 C)。这 A、B、C 和 D 的时间总和也就是完成一组集并移到下一组集的组集与组集间的间隔时间(图 4-5)。

③组集内=C+D。处理组集内信息的反应时间是处理组集内各个特点单元间的潜在反应时间。组集单元是构成组集的特点构件。由于认知系统必得在绘完一个单元之后察知绘出的结果,以便决定回收下一个单元构件(需时 D),而后由心智扫描影像决定下一待绘单元构件的大小(需时 C)。因此,处理

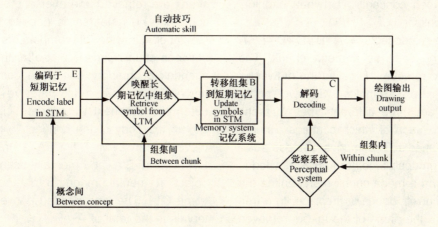

图 4-5　心智进行过程图例
Figure 4-5　Mental processing diagram

mat. The cognitive mechanism for decoding is scanning, which is to mentally scan the object in imagery to determine the dimension of the object. The **scanning** time is C.

(5) After the mental imagery is scanned and drawing actions are (executed and) completed, the perceptual system will perceive the completed features on the drawing to determine the next object to be retrieved. The time needed for **perception** is D. Both decoding and perception are recursive until contents in STM are exhausted. Only when there are no symbols left in STM will the next symbol from LTM be retrieved and the next cycle of processing be started.

The entire sequence of imaging and drawing a picture and the stages in which they are performed is paramount. This means that the second stage in the process cannot be executed without first satisfying the previous condition. The reaction times are the real times needed to perform mental tasks, and a distinct amount of time is needed to complete each action. If a task requires the involvement of several mental operations, its reaction time is the total time for executing these operations. Thus, there are three categories of reaction time in operating mental images: between concepts, between chunks, and within chunks.

① **Between concepts** = A+B+C+D+E. A concept can be seen as a mega chunk. The time needed to retrieve it is called the between-concept time. For example, window, door, and fireplace are three different types of architectural forms, each of which contains a number of sub-elements. The images of these sub-elements are grouped together in the knowledge structure to form a mega chunk representing the concept of window, door, or fireplace. In a free recall process, the name of the to-be-retrieved image must be memorized first. This is done by encoding its verbal symbol in STM, which takes time E. To evoke the image chunk from LTM that matches the label in STM takes time A, and to transfer it into STM takes time B. The activated image chunk in STM is an abstract isomorphic to a real object, and it takes time C to scan through this image to determine the size of the to-be-drawn object. Then a drawing starts. After the drawing action is completed, the perceptual system perceives the presented features in the drawing context to determine the next action to be proceeded. This perception takes time D. If the next action is to retrieve a concept, then a new label is needed to serve as a stimulus cue that must be rehearsed to encode the label in STM. The whole sequences of recall repeats. Therefore, the **between-concept reaction time** is defined as the time needed to perceive the situation (D), encode the label in STM (E), evoke the chunk in LTM (A), transfer the chunk to STM (B), decode it (C), up to the point of starting a drawing, and it is the sum of D, E, A, B, and C (Figure 4-5).

② **Between chunks** = A + B + C + D. After a concept has been retrieved, the chunks contained in the concept are triggered. While handling the first chunk, components inside that chunk are retrieved one after another by activation, and no rehearsal is needed for encoding the label. Drawing of components starts. The between chunks time is that after the drawing of a chunk of imagery has been completed, the perception system must perceive the results, determine the next appropriate chunk (time D) in the body of the concept to proceed, retrieve it from LTM (time A), transfer it into STM (time B), and mentally scan through the object in mental image to determine its size (time C). The time needed **between chunk** intervals is the total of A, B, C, and D (Figure 4-5).

③ **Within chunk** = C + D. The within chunk

组集内信息所需的时间即是 C 和 D 的总和（图 4-5）。所有在心智进行过程（图 4-5）中明说明的心智反应时间，可综合详列于表 4-1 中。

心智过程中反应时间的总表　　　表 4-1

行　　动	总共的反应时间
由长期记忆回收	A
转　　移	B
解　　码	C
察　　觉	D
编　　码	E
组 集 内	C+D
组 集 间	A+B+C+D
概 念 间	A+B+C+D+E

理论假说：

图 4-5 即以图解显示由回记一份影像到完成绘出该影像的全部心智过程。这图解也提供一些机会由探测组集间和组集内单元构件的反应时间，来发掘可能的知识结构。三个理论假说也依此设立。一系列心理实验也随后进行，以验证这些假说。

假说一：设计专家曾经历过许多设计，也因此应累积并拥有更多的设计知识。相对地，设计专家应比外行人储存更多的建筑符号（组集）。因为越多组集和内涵的信息会提供更多的造型数据和几何形体。因此，设计专家丰富的内在呈现表征和相对的心智影像会促进外在表征，也会绘得比外行人更详细也更丰富。这一假说和却斯与司马（1973，1974）的西洋棋实验，瑞特曼（1976）的围棋实验（见附记 2），以及埃肯的探索建筑符号实验（1978）相符。

假说二：设计专家的知识本体应该在数量上及组集内涵上大于外行设计者。组集并非任意组合成群，它们应以某些方式集合并结，因而数据的回记会更有效、更快速。**建筑机能应是建筑知识本体中组合信息的方式**。设计专家应有更大的组集群。因此它们拥有的知识本体应更能提供证据，证明比外行人有更强烈以建筑机能组合知识的倾向。

假说三，由记忆中回取组集的反应时间，可用于探查两组不同的知识表征有何不同的认知绩效。**如果设计专家储存信息有协调一致的方法，则回收信息的能力将效率更高，而且反应时间也更短**。如此，专家应花更少时间由长期记忆中回记数据，即图 4-5 中的变数 A。

心理实验：

一个在实验室进行的心理实验被设计来验证上述三个假说。这实验要求受测者心智作出在不同地方的壁炉影像。由于壁炉在民居建筑中是个普遍而且显著的特征。所有受测者都应可轻易制造出这些影像。制造影像的方法无任何限制，受测者可回记任何曾经看过的影像。但回记出的影像是被要求以线条绘于纸上。受测者来自卡内基梅隆大学小区，并分成专家和外行新手两组。专家组有四位，有执业建筑师，并且也有建筑系教授。外行新手组有两位，是随机抽样选取，而且属于非建筑专业。马克笔和 8.5 英寸乘 11 英寸的白纸也在实验中提供，以便绘图之用。

实验程序及方法：

实验使用的是重复处理法。受测者被要求想像并且画出五个不同种类的壁炉，包括：位于旅馆大厅、位于度假中心的休闲室、位于民居客厅、一个现代壁炉以及一个古典壁炉（以符码 1、2、3、4、5 各自代表）。受测者会先要求想像一个特定种类的壁炉，然后快速准确而且详细地以笔绘出。绘完之后，即以**口语出声法**（见附记 3）把全部画出的元素从小到大以圆圈勾勒，并且写下这些单元的名字。这部分与瑞特曼要求她的受测者圈出**围棋**部分以确认围棋团块的实验相似（1976，1980）。在这个阶段，如果受测者无法圈出单元或无法定名字，则要求以口语描述该单元，因为绘画能力是**外在变量**（见附记

reaction times are latencies among chunk components which are the features that constitute the chunk. Because the system must perceive the drawing results after a component of features is drawn to determine the next component to be retrieved (time D), and mentally scan through the item in mental imagery to determine its to be drawn size (time C). The time needed for **within chunk** is the sum of C and D (Figure 4-5). A summary of all reaction times specified in the processing model is listed and diagrammed in Table 4-1.

A summary of reaction times specified in the model
Table 4-1

Action	Total reaction time
Retrieve from LTM	A
Transferring	B
Decoding	C
Perceiving	D
Encoding	E
Within chunk	C+D
Between chunk	A+B+C+D
Between concept	A+B+C+D+E

Hypotheses:

The diagram shown on Figure 4-5 explains the cognitive process from recalling an image up to the completion of the image drawing. This provides opportunities to discover the potential structure through the exploration of reaction times among chunks and between chunks. Three hypotheses had been addressed. A series of psychological experiments were conducted to test the hypotheses.

Hypothesis one: *Expert designers experience more designs and thus, accumulate and possess more design knowledge.* Accordingly, they also should store more architectural symbols (chunks) in their knowledge than do novice designers. Because more chunks and their embedded information provide more forms and, possible, more geometric shapes; expert's abundant internal representations and corresponding mental image enrich their external representations, and they can externally draw more than novices. The hypothesis is parallel to chess experiments by Chase & Simon (1973, 1974), Go (see note 2) experiments by Reitman (1976), and exploration of architectural symbols by Akin (1978).

Hypothesis two: Expert designer's knowledge base should be quantitatively larger and hold more chunks than that of a novice counterpart. Chunks are not randomly clustered in groups. There should be means that bind chunks together so that information retrieval can be made rapidly and efficiently. *Architectural function could be the means that organize architectural information in the knowledge base.* Expert designers have larger groups of chunks, so their knowledge bases should provide evidence of a stronger tendency to utilize function to organize architectural knowledge than that shown by the novice designers.

Hypothesis three: Reaction times for retrieving chunks from memory are used to detect two groups' cognitive performance under the effect of different knowledge representations. *If expert designers store information in a more coherent fashion, they should be able to retrieve information efficiently, and their reaction time should be accordingly shorter.* Thus, expert designers should retrieve chunks in less time from LTM, the variable of A in Figure 4-5.

Psychological experiments:

A laboratory experiment was designed to

4)之一,它可借口语描述绘出的特征将其平衡化。最后,受测者被要求以树状结构写下所画壁炉的所有单元体系。第一及第二阶段重复五次完成五个不同壁炉。第三阶段的树状结构是在五次实验处理完之后,再写出这构架。目的是探讨受测者自定的典型知识结构。绘图时间无限制,整个过程被录像机以录像带(VCR)全程记录下来。

变量及衡量方法:

有三种衡量法度可用来探测知识结构的本质。一是停顿时间,停顿时间即是测量在两个连续绘图动作中间的延迟并转换动作的时间。度量方法是将录像带减速,计算影带中影片的格数。依照录像机(VCR)1秒钟放映30格的规格,下列公式即可将影片格数换算成正确的实际时间:

停顿时间=每格 0.033 秒×格数

在这实验中,由录像带计算停顿时间的方法,是由受测者在绘图的动作停止那一刻算起,而不是由笔离开图纸算起。停顿时间也被分成三个类别:(1)组集内时间:这是完成一个单元特征的连续动作之反应时间;(2)组集间时间:这是在第一个特征完成,而后开始第二个特征之间的停顿时间,亦即分开两特征间的空档期;(3)观念间时间:在两个不同建筑形态(元素)之间的停顿时间。

第二个衡量法度是内在表征的数目。一个单元的内在表征,是受测者在画上勾出一圈,并附上一个符号的单位,也被算记为一个组集,由此,单位数可研究出受测者知识库中所存知识组集的数目。内在表征有三个来源:(1)口语描述中的符号(**文字码**);(2)图中画出的特征(**视觉码**);(3)特征单元(组集单元)间的停顿时间。主要课题是把口语中符号和图中绘出的特征相配对。一个图中特征是一个有特别配置的具体形体。如果符号和图是一对一配,则符号可看成是象征知识结构里的组集。如果一个符号配上多个特征,或全不配,则停顿时间即用来确定组集的界限。长期停顿时间即宣示两个组集是

不同的表征,而应被认定是两个不同组集。

第三个衡量法度是机能关系。机能关系视连续两个组集出现的关系而定。两组集是否连续即由 A(X、Y)的命题决定。A 代表建筑关系,X 及 Y 代表两连续的组集。依理而言,每一组集应有许多关系可与其他组集多方联系。但在回记的过程里,组集的激化是在长程记忆中由活跃的组集蔓延到其他部位,但能被激化的次部位,显然是最靠近该部位或与该部位有最强联系者才能体现的(Collins & Loftus, 1975; Anderson, 1980)。虽然在实验分析中不可能把整个知识架构中的联系全部都能辨识得出,但仍可能认识到强烈而且有直接联系的组集。

实验结果:

实验结果非常有趣。第一位著名设计专家的图十分概略,富有写生的意趣。每根线条都有数根笔触。每个元素都是雏形抽象画,而且缺乏细节数据,边线也没明确或肯定的定义。事实上,这受测者是在做设计而非回记已记住的影像(图4-6)。第二位设计专家则用剖面图来显示影像。形在其机能被确定之后才浮现。这些图也建议了受测者是在做设计而非回想影像(图4-7)。因此这两组图都不符合研究方法而排除在资料分析中。其他合适的外行新手A、B和设计专家A、B绘出的图形则依序列于图4-8~图4-11中。整体而言,两位设计专家和两位外行新手作出20张图,产生下列的实验结果。

文字码与**视觉码**:在大部分的图中,圆圈代表绘出特征的文字符号,并且能和画出的特征相符合。也有没法配合的情况发生,这是由于受测者无法将该特征定名,但能以口语描述该物。例如外行新手B即以"火炉嘴"代表"壁炉台架",以"嘴内"代表"壁炉膛"。这显示几乎每个影像特征单元都会有一个可说的文字码相连。

内部表征的数量:出现在图中的单元特征都被认为是影像的部分,也被看作是存在于知识体中代表心智影像的知识组集。任何特征(或符号)被同

test the hypotheses. The experiment required subjects to generate images of fireplaces in different settings. Because a fireplace is a common and prominent feature in many residential buildings, subjects should not have any difficulty generating such images. The method of image generation was not restricted and subjects could recall any image they had seen before. The recalled images were required to be presented in line drawings on paper. Subjects were from the Carnegie Mellon University community. They were divided into novice and expert groups. Four subjects in the expert group were licensed architects and faculty members in the architecture department while the experiments were conducted. The two novice designers were randomly selected students with no architectural design background. Magic Markers and standard 8.5 by 11 inch white paper were provided in the laboratory for drawings.

Experimental procedures and methods:

Experiments were repeated treatment method. Subjects were asked to generate and draw five different kinds of fireplaces: for a hotel lobby, a resort center recreation room, a family living room, a contemporary fireplace, and a classical fireplace (coded as numbers 1, 2, 3, 4, and 5, respectively). They were asked first to form an image of a particular type of fireplace and drew it with as much detail as possible and as quickly as possible. After the drawing was completed, subjects were secondly asked to **think out loud** (see note 3) while identifying chunks by circling and labeling the drawn element from small pieces to larger ones until the whole drawing was labeled. This stage is similar to Reitman's method of asking subjects to circle the recalled **Go** units to confirm the unit of chunk in her Go experiments (Reitman, 1976; 1980). At this stage, if the subjects could not circle and label a specific feature, they were asked to describe it verbally. Because drawing ability is an **extraneous variable** (see note 4), it can be balanced out by the use of verbal descriptions of features. In the last stage, the subjects were required to write down the hierarchy of all elements of a fireplace in a tree-like structure. The first and second stage were repeated five times for the five different types of fireplaces. The third stage was done at the end of the experiment to test prototypical knowledge structure. The drawing time was not limited, the entire drawing process was videotape recorded in the video cassette recorder (VCR) format.

Variables and measurements:

Three measurements were used to explore the nature of knowledge structure. The first one was **pause time**, which was defined as the time elapsed between two successive drawing actions. It was measured by slowing down the videotape and counting the number of frames in the tape. According to the VCR video format, one second contained 30 frames, so the counts of frames were converted into real time by the formula of:

Pause time = 0.033 second per frame × number of frames

In this experiment, the moment the drawing action ceased, rather than the moment the pen left the paper, was selected as the point to start measuring pause time. Furthermore, the pause times were divided into three categories of: (1) within chunk time: the reaction time between successive drawing actions for completing a feature; (2) between chunks time: the pause between finishing the first and starting the second feature, or the time of separating two features; and (3) between concepts time: the pause between two different types of architectural forms (elements).

The second measurement is the **number of internal representation**. A unit of internal representation compatible with the label name that

一受测者在不同的图中重复绘出的，只能算是一个特征单元，遑论其出现的频率次数。整个出现的组集和受测者写出的符号列于表 4-1。专家组的组集（平均值=73）和符号（平均值=40.5）大于外行新手的组集（平均值=44.5）及符号（平均值=19），差异是显著有效的（组集的概率值<0.05，符号的概率值<0.005）。

树状体系层次：所有受测者写出的树状体系结构显示出内在心智表征可以系统地安排出。专家的树包含更多层次的节点和更多信息，然而新手外行则因为无法提供更细部的单元，因此比专家组有更少的特征绘出。

影像组合：在这实验中，受测者完成位于旅馆大厅、度假中心的休闲室以及住宅客厅的三个壁炉影像之后，他们被要求想像并画出一个现代和一个古典的壁炉。结果显示外行新手 A 的度假中心壁炉相似于现代壁炉，而且其他三个壁炉的造型及用材几乎都相似。外行新手 B 的度假中心壁炉相似于现代壁炉，而且旅馆大厅壁炉神似于古典壁炉。相反的，专家组的壁炉画在造型及用材上几乎都不同。例如，专家 B 特别用"梁楣"或"飞檐"来形成"柱顶过梁"。这些都在他的口语中描述成构成"壁炉顶架"的元素。他说明了这种元素是采自柱梁结构。在他的图中，所有"壁炉顶架"、"壁炉台架"和"壁炉上方横楣"都是柱梁的缩版。这种造型出现在他的旅馆大厅壁炉和古典壁炉两例，但细部稍有变动。相似的情况发生在专家 A 的图中，她也用"饰带"和"齿状装饰"构成"壁炉顶架"，形成一

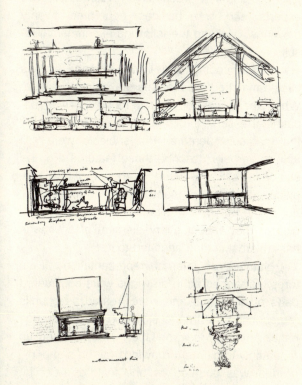

图 4-6　设计专家一的图和其提供的树状体系结构
Figure 4-6　Drawings and the hierarchical structure provided by expert designer 1

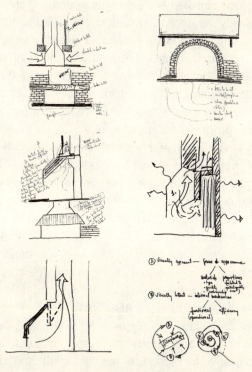

图 4-7　设计专家二的图和其提供的树状体系结构
Figure 4-7　Drawings and the hierarchical structure provided by expert designer 2

subjects circled in drawings was measured as a chunk to study the number of chunks existing in each subject's knowledge base. There were three sources for determining an internal representation: (1) labels from verbal description (**verbal coding**), (2) features shown in drawings (**visual coding**), and (3) pause time between features (chunk unit). The primary task was to pair labels in verbal descriptions with features in the drawings. A feature was a shape with some specific configuration. When there was a one-to-one match, that label was considered to be the symbol representing the chunk in the knowledge structure. If a label matched more than one drawn feature, or there was no match at all, then pause times were used to verify the boundary between chunks. A long pause signified two elements appearing consecutively as having two different representations and therefore being treated as two chunks.

The third measurement is *functional relation*. Functional relations were detected by the relationship between two successive chunks. The link joining two chunks was expressed in the propositional notation of A (X, Y), where A stood for an architectural relations, and X and Y were labels of two successive chunks. Theoretically, there were a number of links connecting one chunk with others. In the retrieving process, however, activation spread from active portions into other portions of LTM so that the next chunk to be retrieved was the one most closely related to that chunk or the one that had the strongest link (Collins & Loftus, 1975; Anderson, 1980). Although it was impossible to completely identify all links in the knowledge structure through the analysis, it, yet, was possible to recognize the strongest and direct link between chunks.

Experimental results:

Results obtained in this experiment were very interesting. One expert's drawing was very sketchy, with each line repeated by multiple strokes. Elements were primitive shapes drawn abstractly without having detailed information. Edges were not precisely defined. The subject was designing new forms instead of recalling stored images (see Figure 4–6). Another expert used section drawings to show images; forms emerged only after the function of forms was determined. Drawings suggest that this subject was also conducting design rather than image retrieval (see Figure 4–7). Thus, these two sets of data did not fit the research methods and were excluded for the data analysis. Qualified drawings done by novice A, B and expert A, B were shown on Figure 4–8, 4–9, 4–10 and 4–11 respectively. Altogether, two experts and two novices generated twenty drawings, yielding the following experimental results.

Verbal coding and visual coding: Labels circled matched the features drawn in most data. When there was no match found, it was because the subject did not know the name for that particular feature. But subjects were still able to verbally describe these features. For instance, novice B used "mouth of the fireplace" to stand for mantelpiece and the "interior of the mouth" to symbolize firebox. This indicates that almost every element of an image has a verbal coding attached to it.

Number of internal representation: Features that appeared in drawings were considered parts of an image and were treated as chunks representing the image stored in the knowledge base. Any feature (or label) that showed up repeatedly in different drawings by the same individual was counted as one features (or label), regardless of the frequency of its appearance. The total numbers of chunks and labels generated are shown in Table 1. The expert

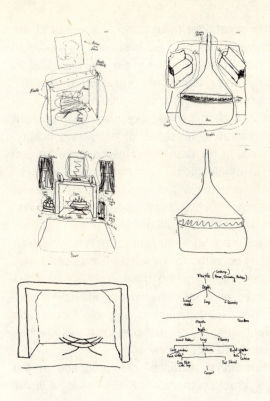

图 4-8 外行新手 A 的图和其提供的树状体系结构
Figure 4-8 Drawings and the hierarchical structure provided by novice designer A

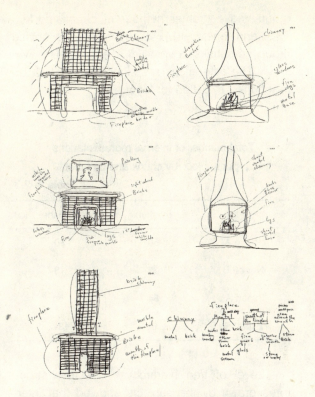

图 4-9 外行新手 B 的图和其提供的树状体系结构
Figure 4-9 Drawings and the hierarchical structure provided by novice designer B

种古典主义的特色。这些例子说明了专家设计师有能力把不同的影像合并，或将旧影像（组集）重新组织，创造出新影像。

机能关系：依照影像结构模式（图 4-2）所提的概念，记忆中的视觉组集由联想串成。当信息被扣取时，具有强烈联系的组集即逐一地以线性秩序被激化回收。组集回收的顺序配合外显于徒手绘图上的呈现都可透出记忆中组集互连的特色和关系。基于此理，观察回记顺序会提供契机来了解组集在记忆中的组织情况。图 4-13 即以图形示例外行新手和专家设计的两个影像回记程序。两个图形中的虚点线表示节点间的互连关系，粗实线带箭头即示回记的次序，并以数字标出回记的前后次序过程。

于 67 页所提的 A（X，Y）之命题，即用来决定两个组集间是否肯定有建筑机能相连之判断方法。其中 X 是一个符号个体（或图上绘出的单元），Y 是随后的另一符号个体（或随后绘出的单元）。两符号个体间的关系 A 如满足下列中的一项，即可宣告此两单元有建筑机能的联结：(1) 结构联系（包括材料）；(2) 空间联系；(3) 约定俗成的习俗联系（即 Y 的外观是由其使用而决定，这使用是为满足 X 的目的，反之亦然）；(4) 美观联系（这包括为装饰而用的修饰品）。表 4-3 是清单列出所有出现在 20 张图的壁炉单元间之可能的机能联系。

由收集组集出现的前后顺序数据，并由表 4-3 清理的关系作判定，所有单元与单元间的机能联系

group had a greater number of chunks (M= 73) and labels (M=40.5) than the novice group's chunks (M=44.5) and labels (M= 19). The differences are significant (P<0.05 for chunk, p<0.005 for label).

Total number of internal representations per subject, by chunk and by label

Table 4-2

	By chunk	By label
Novice A	50	20
Novice B	39	18
Expert A	84	50
Expert B	62	31

Level of tree hierarchy: The tree structures drawn by the subjects showed that internal representation can be arranged systematically. Experts' trees contained more levels of nodes and more information, whereas the novices were unable to provide details of the elements and, thus, showed fewer features than experts.

Image composition: In this experiment, after subjects completed three different images of fireplaces for a hotel lobby, for a recreation room in a resort center, and for a living room in a house, they were asked to generate and draw a contemporary and a classical fireplace. Results showed that for novice A, the drawing for a resort center was similar to the contemporary one and the other three drawings were similar in shape and in the use of materials. For novice B, the fireplace for a resort center was similar to the contemporary one, and the hotel lobby fireplace looked like the classical one. In contrast, for the expert designer group, drawings of the fireplaces are all different in shapes as well as in materials used. For example, expert B had specified using a lintel or cornice to form the architrave, which constitutes the mantelshelf. In his verbal descriptions, he indicated that such elements are adaptations of beam and column structure. In his drawings, the composition and form of mantelshelf, mantelltree, and mantelpiece were miniatures of beam and column structures. Such a form appeared in the fireplace for the hotel lobby and in the classical fireplace, but with changes of details. A similar situation occurred for expert A, who used fascia and dentil to form the mantelshelf, creating a character of classicism. These examples suggest that expert designers are able to combine different images or to reorganize old images (chunks) to create new ones.

Functional relations: As proposed in the image structure model (Figure 4-2), visual chunks in memory are linked by association. While information is being retrieved, chunks in strong links are activated and recalled one after another linearly. The sequences of retrieving chunks and presenting them on drawings should reveal the characteristics and the relationship that link chunks in memory. Based on this theory, observations on the retrieval sequences should provide opportunities for understanding the organization of chunks in memory. Figure 4-13 shows graphical examples of a novice designer's and an expert designer's retrieving sequences. In both diagrams, dotted lines stand for links between nodes, and heavy lines with arrows show the retrieval sequence which was marked by numeric numbers.

To determine if a link between two chunks has an architecturally functional connection, the proposition A (X, Y) discussed on page 68, where X is a symbol and Y is the one that follows,

图 4-10 设计专家 A 的图和其提供的树状体系结构
Figure 4-10 Drawings and the hierarchical structure provided by expert designer A

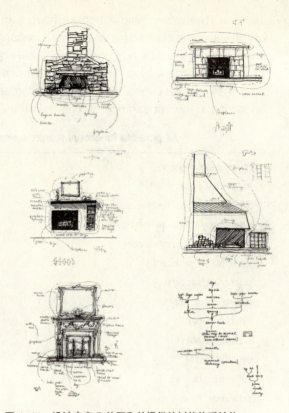

图 4-11 设计专家 B 的图和其提供的树状体系结构
Figure 4-11 Drawings and the hierarchical structure provided by expert designer B

总结于表 4-4。在这表中，单元相连总数是由回记组集的间隔数而定。例如，新手外行 A 的第一图（图 4-8）有 8 个组集、7 个间隔相连（在此例中，三个炉材原木只能算是一个炉材原木单元，重复出现三次）。这 7 个连接，参考表 4-3，显出只有两个有机能相连。平均而言，新手的机能相连率是 50%（新手 A）及 54%（新手 B），专家则是 72%（专家 A）和 79%（专家 B）。这结果显示专家的知识网络最少有 72% 以机能互联。尤其是专家 A 的第五图，为全部图形中单元出现最多（总数 30）的一图，而且 24 个组集中有 23 个连续出现是有机能关系互动的。

组集回收的参数：实验中组集内的全部停顿次数是 944 次，组集间的停顿次数则是 214 次。受测者平均的组集内和组集间停顿次数列于表 4-5。比较专家和外行，两组间的组集内（概率值<0.0001）和组集间（概率值<0.0001）反应时间有非常显著的差异。如预期的，专家设计者有比外行新手更短的组集内及组集间之停顿时间。

如心智进行过程图（图 4-5）所示，组集内的停顿时间包括解码（C）和察觉（D）两个动作。组集间的停顿期则包括回记（A）、转移（B）、解码（C）和察觉（D）四个动作。如果由组集间减去组集内时间，剩下的即是回记（A）和转移（B）的净值。丹色律（1969）曾测出转移时间（B）是 0.3 秒。故由 A+B 公式中去掉 B（0.3），可得出大约的 A 值（表

is utilized. The relationship A between two symbols of X and Y is claimed to have some functional connection if it satisfies any one of the following: (1) structural connection (this includes materials); (2) spatial connection; (3) customary connection (the appearance of Y is determined by its usage, which serves X purposes, or vice versa); (4) artistic connection (this includes the ornamental elements for decoration purposes). A list of all possibly functional connections among fireplace features that appeared in 20 drawings is given in Table 4–3.

All possible functional relations among features in a fireplace Table 4–3

(1) Structural connection	(2) Spatial connection
sit (chimney flue, mantel)	above(mantel, firebox)
sit(chimney hood, mantelpiece)	above(lintel, firebox)
sit(chimney hood, chimney breast)	above(fascia, dentil)
sit(mantelshelf, mantelpiece)	below(firebox, lintel)
sit(mantelpiece, hearth)	below(floor, glass screen)
sit(cornice, architrave)	embrace(mantelpiece, firebox)
sit(architrave, capital)	embrace(mantelpiece, mesh screen)
sit(shaft, base)	embrace(firebox, glass screen)
sit(lintel, jamb)	embrace(firebox, andiron)
sit(jamb, base)	inside(glass screen, firebox)
sit(firebox, hearth)	inside(meshscreen, firebox)
sit(firebox, floor)	inside(firebox, mantelpiece)
sit(hearth, floor)	within(log, firebox)
sit(nosing, riser)	
sit(riser, floor)	(3) Customary connection
sit(keystone, firebox frame)	hold(andiron, logs)
support(floor, mantel)	hold(candlestick, candle)
support(floor, andiron)	contain(vase, flowers)
support(floor, mesh screen)	generate(log, fire)
support(floor, firebox)	
support(hearth, mesh screen)	(4) Artistic connection
support(cornice, lintel)	decoration(chimney breast, mirror)
support(mantelpiece, chimney breast)	decoration(chimney breast, picture)
support(mantelpiece, mantelshelf)	decoration(mantelpiece, classical swags)
support(mantel, chimney flue)	decoration(mantelshelf, fascia)
fasten(screen, screen frame)	decoration(firebox, keystone)
construct(mantelpiece, rustication)	decoration(painting, frame)
construct(mantelpiece, natural stone)	
construct(chimney breast, brick)	

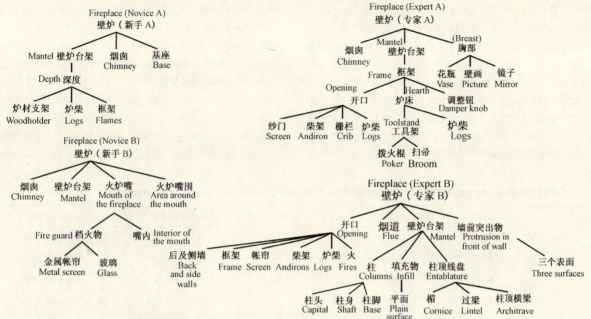

图 4-12 设计专家（右图）和外行新手（左图）写出的树状结构
Figure 4-12　Written tree structures by experts （right parts） and novices （left parts）

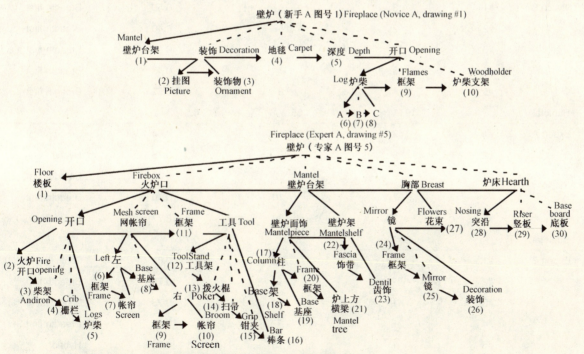

图 4-13 外行新手 A（图号 1）和设计专家 A（图号 5）的回记程序
Figure 4-13　Retrieval sequence of novice A （drawing #1） and expert A （drawing #5）

Based on the collected sequential appearance data and evaluating the obtained relations in Table 4-3, results of the functional links among features are shown in Table 4-4. In this table, the total number of links in a drawing is determined by the number of intervals between retrieved chunks. For instance, the total number of chunks in novice A drawing #1 (Figure 4-8) is eight, and the number of links is seven. (In this example, three logs are counted as one log chunk repeated three times.) These seven links, compared with Table 4-3, have two links that are functional. On average, the functional links for novices are 50% (novice A) and 54% (novice B), whereas functional linkage by the experts are 72% (expert A) and 79% (expert B). This result suggests that, in the experts' knowledge network, at least 72% of the

The number and probability of functional links between chunks and their search methods   Table 4-4

	Drawing	Functional link	Total links	Probability	Search strategy
Novice A	#1	2	7	0.29	Breadth first
	#2	4	7	0.57	Breadth first
	#3	6	12	0.50	Breadth first
	#4	2	4	0.50	Breadth first
	#5	3	4	0.75	Breadth first
Subtotal		17	34	0.50	
Novice B	#1	3	6	0.50	Breadth first
	#2	2	5	0.40	Breadth first
	#3	7	11	0.64	Breadth first
	#4	3	7	0.43	Breadth first
	#5	4	6	0.67	Breadth first
Subtotal		19	35	0.54	
Expert A	#1	7	11	0.64	Breadth first
	#2	6	8	0.75	Hybrid
	#3	9	13	0.69	Breadth first
	#4	7	10	0.70	Breadth first
	#5	23	30	0.77	Depth first (90%)
Subtotal		52	72	0.72	
Expert B	#1	13	16	0.81	Depth first (88%)
	#2	7	9	0.78	Breadth first
	#3	11	14	0.78	Depth first (91%)
	#4	6	7	0.86	Breadth first
	#5	11	15	0.73	Hybrid
Subtotal		48	61	0.79	

4-5），也显出专家设计者都有较短的回记时间。这较短的时间表示专家由长期记忆中能快速回收数据，一方面是基于有效的搜寻策略，或另一方面是有较好的知识架构。但在较短的译码（C）和察觉（D）时间里，无法说明究竟是哪个因素（C 或 D，或两者）扮演了重要的角色。

图 4-14 是受测者组集内和组集间停顿时间关系图。受测者有较长的组集间停顿期，则其组集内的停顿期也较长。这结果显示了反应时间和设计能力成反比例：较佳的设计能力具备较清楚的视觉码，并花费较少的回收时间。在此实验中，专家 B 有多于专家 A 五年的建筑设计专业经验。可预测的是如果把受测人数增加，则组集内和组集间的反应时间会和设计能力保有相似的斜线走向。

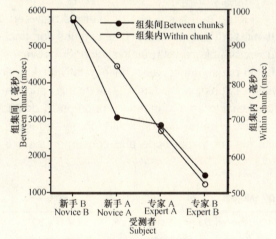

图 4-14　受测者组集内和组集间停顿时间关系图
Figure 4-14　Relation between within-chunk and between-chunk pause time, by subject

关于概念间的停顿时间，实验并没带出足够数据做进一步分析。但概念间的停顿比组集内和组集间停顿时间要长得多，这可由窗和壁炉之间看得出。窗和壁炉并不属于相同的组集，因此两个组集应算是两个不同的概念组集。外行新手 A 和专家 A 两人都花了长于 10 秒的时间回收窗户的影像。

扫描时间：扫描时间是扫描心智影像，决定物体真正尺度的时间。要测定扫描时间，首先得假定存在心中的比例尺用在每张图都与真正实际尺度无关，而且对所有受测者而言都是恒定的。循此假设，同一图中的两组线（受测者绘任何物体所用的两条连续线）即被选出做研究。在绘这些线时，有两种心智决定必得考虑。第一个决定（种类一）是多远之外开始画第二根线。这一类的第二根线通常与第一根线平行，而且有相同的长度。第二个决定（种类二）是确定第二根线的长度。所有此类的第二根线与第一根线垂直。理论上，当第一根线已完成，受测者必须扫描心智影像的大小，再决定在多远之外来定位这第二根线；或多长来决定这根线。因此，需要一些反应时间。

受测者的内在表征之总数包括符号和组集　　表 4-2

	组集数	符号数
新手 A	50	20
新手 B	39	18
专家 A	84	50
专家 B	62	31

这两个扫描决定可提供一个机会，观察物体大小和反应时间的关系。表 4-6 列出两个例子。在例子中，停顿时期的长短正与所绘物体大小相配。在所有数据里，种类一有 9 个并行线案例，其中 7 个满足这倾向。种类二则有 6 个垂直案例，其中三个显示出反应时间和扫描距离成正比。这结果也部分支持考斯林等（1978）的发现，即扫描影像中两物体的时间与线性距离有关。也有可能置疑说本实验中的扫描时间内包含手的移动，但手的移动和扫描过程同时进行，而且手中笔的移动可当成是认知收受系统的一个指针，提供空间数据与记忆中的影像相配合来启发画图的动作。

搜寻信息的策略：搜寻策略是人工智能学科中主要的研究重点。在认知科学中，它最早被纽韦尔和司马（1972）系统地探讨过。在本实验中，三种自然的搜寻策略由数据中得到证实。第一种是**深度**

links are based on architectural functions. Specifically, in drawing #5 by expert A, the drawing has the largest number of features (total of 30) among all drawings, and 23 links between 24 features are functional.

Chunk Retrieval Parameters: The total number of within-chunk pauses observed was 944, and 214 pauses were observed between chunks. The means of between-chunk and within-chunk pause times for each subjects are shown in Table 4-5. The comparison between the expert and novice groups showed highly significant difference in reaction times within chunk ($F(1, 934) = 33.65$, $p < 0.0001$) and between chunks ($F(1, 204) = 27.33$, $p < 0.0001$). As expected, expert designers had shorter pause times within chunk and between chunks.

As described in the mental processing model, the within-chunk pause times consist of the factors of decoding (denoted as C) and perceiving (denoted as D). Because the between-chunks pause is the sum of retrieving (A), transferring (B), decoding (C), and perceiving (D), by subtracting the latencies within chunks from those between chunks, one can obtain the value for A + B. As mentioned before, B has been measured by Dansereau (1969) to be 0.3 seconds. By subtracting 0.3 seconds from A + B, A (the time for retrieving) can be estimated (Table 2-5). From the values displayed in Table 2-5, it can be seen that the two experts had shorter reaction times. A smaller A means that experts were faster in retrieving material from LTM, which may be the result either of having a more efficient strategy or of having a better-structured knowledge representation. Less time also was spent by the experts to do the decoding (C) and perceiving (D) together, but the findings do not tell which factor (C or D, or both) plays a critical role in the process.

Figure 4-14 is a plot of average between-chunk and within-chunk pause times for each subject. Subjects with relatively longer between-chunk pause times also have relatively longer within-chunk pause times. The results suggest that reaction time stands in proportion to design expertise—better design ability comes equipped with clearer visual code and consumes less re-

	Summary of pause times, by subject			Table 4-5
	Between chunk = A+B+C+D (msec)	Within chunk = C+D (msec)	Between chunk - Within chunk = A+B (msec)	A + B - 300 = A (msec)
Novice A	3070 (43 observations)	846 (184 observations)	2224	1924
Novice B	5702 (34 observations)	978 (187 observations)	4724	4424
Expert A	2842 (81 observations)	668 (360 observations)	2174	1874
Expert B	1469 (56 observations)	525 (213 observations)	944	644
Total	214 observations	(944 observations)		

优先搜寻。在设计中，它要求先把每一个物体的边框（或轮廓）加细部全部画成后再进行下一个物体。第二种是**广度优先搜寻**，这意味着把每一个物体的边框轮廓先完成再加个别的细部。在此实验中，探测搜寻策略的方法是把出现在20张画中所有的壁炉单元，按机能连接组成一个假定的知识表征（图4-15）。然后追踪每张图画出单元的次序来鉴定每个受测者所使用的认知搜寻策略。在拟出的知识表征中，楼板的组集位置是假定在炉床台的大组集下。但如画里没有炉床台，或这画出的壁炉并没有炉床台，则楼板就移位于壁炉膛的部分组集里。

图4-16是专家A于实验序列图4-10画现代壁炉的搜寻策略结果。绘这图的搜寻形态显示出大部分的步骤运行，是将整个组集完成后才进行下一组集。比方说，此受测者会依序把壁炉膛完成后再画壁炉台架。在所有30个绘图步骤中，有27个是深度优先搜寻，其他三个步骤属广度优先法：即在画楼板后（即动作1到2），画齿状装饰后（即动作24到25），以及画花朵之后（即动作27到28）。因为90%的步骤是深度优先法，因此这位专家受测者可以说是用此法来撷取信息的。

图4-17是另一图例，这是外行新手B画一家居客厅壁炉的顺序。这搜寻步骤表示出先画壁炉台架的轮廓，再画防火网帘的外廓，最后是壁炉膛的外形。之后，再把所有组集的细部依序加入。这形态就显示出了广度优先的策略。

混合深度优先及广度优先的策略也被两位专家受测者使用。图4-18即是专家B用深度优先法画壁

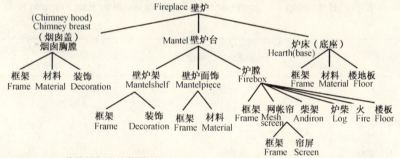

图4-15 草拟的知识结构模型
Figure 4-15 Prototypical knowledge structure model

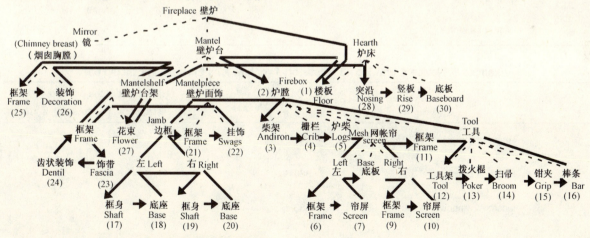

图4-16 专家A于图4-4中所用的深度优先搜寻策略
Figure 4-16 Depth-first search strategy of expert A, drawing #4-4

trieval time. In this experiment, expert B has five more years of professional architecture design experiences than expert A. It is predicted that if the sample sizes of novice and expert groups were to increase, the reaction times both within chunk and between chunks would continue to bear linear relationships to the subjects' design expertise.

Regarding the pause time between concepts, there is not enough data for exploration. Yet, the between-concepts pause is longer than that between chunks or within chunks. This is confirmed by detecting the pause between images of window and fireplace. A window does not belong to the conceptual chunk that is a fireplace, so the connection between the two is considered to be between concepts. It takes longer than 10 seconds for novice A and expert A to retrieve images of windows from memory.

Scanning Time: Scanning time is the time needed to measure the real size of an object by scanning through that object in the mental image. To measure the scanning time, it is assumed that the mental scale for each drawing is independent of the real scale and is constant between subjects. Under this assumption, two sets of lines (two consecutive lines the subjects

Two examples of scanning time   Table 4–6

Example A: Expert B drawing #3 (category one: how far away to start the next line)

Contents	Pause time (msec)	Rank order of the drawing size from large to small
width of breast	1100	1
width of mantel	1000	2
width of rustication	866	3
width of vase	433	4
width of a stone	366	5

Example B: Novice A drawing #3 (category two: how long the next line should be)

Contents	Pause time (msec)	Rank order of the drawing size from large to small
length of mantel	3000	1
length of opening	866	2
length of window	483	3
length of picture	233	4

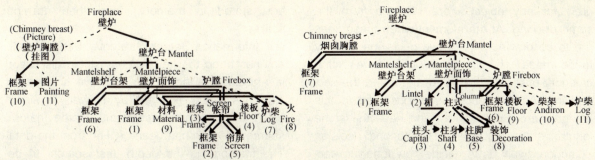

图 4-17 外行新手 B 于图 4-3 中所用的广度优先搜寻策略
Figure 4-17 Breadth-first search strategy of novice B, drawing #4-3

图 4-18 专家 B 于图 4-4 中所用的混合搜寻策略
Figure 4-18 Hybrid search strategy of expert B, drawing #4-4

壁炉单元间所有可能的机能联系清单　　　　　　　　　　　　　　　表 4-3

(1) 结构联系	(2) 空间联系
座(烟囱通道, 壁炉台)	上方(壁炉台, 炉膛)
座(烟囱盖, 壁炉面饰)	上方(楣, 炉膛)
座(烟囱盖, 烟囱胸膛)	上方(饰带, 齿状装饰)
座(壁炉架, 壁炉面饰)	下方(炉膛, 楣)
座(壁炉面饰, 炉床)	下方(楼地板, 玻璃帐屏)
座(飞檐, 横梁)	环绕(壁炉面饰, 炉膛)
座(横梁, 柱头)	环绕(壁炉面饰, 网帐帘)
座(架, 基座)	环绕(炉膛, 玻璃帐屏)
座(楣, 边框)	环绕(炉膛, 柴架)
座(边框, 基座)	内部(玻璃帐屏, 炉膛)
座(炉膛, 炉床)	内部(网帐帘, 炉膛)
座(炉膛, 楼地板)	内部(炉膛, 壁炉面饰)
座(炉床, 楼地板)	在内部(炉柴, 炉膛)
座(突沿, 竖板)	
座(竖板, 楼地板)	(3) 习俗联系
座(拱心石, 炉膛框架)	持有(柴架, 炉柴)
支持(楼地板, 壁炉台)	持有(烛台, 蜡烛)
支持(楼地板, 柴架)	装容(花瓶, 花束)
支持(楼地板, 网帐帘)	产生(炉柴, 火)
支持(楼地板, 炉膛)	
支持(炉床, 网帐帘)	(4) 美观联系
支持(飞檐, 楣)	装饰(烟囱胸膛, 镜)
支持(壁炉面饰, 烟囱胸膛)	装饰(烟囱胸膛, 挂图)
支持(壁炉面饰, 壁炉台架)	装饰(壁炉面饰, 古典挂饰)
支持(壁炉台, 烟囱通道)	装饰(壁炉台架, 饰带)
绑系(帐帘, 帐帘框架)	装饰(炉膛, 拱心石)
构成物(壁炉面饰, 壁面砌石)	装饰(彩绘, 框架)
构成物(壁炉面饰, 自然石)	
构成物(烟囱胸膛, 砖)	

炉顶架、壁炉台架，而后用广度优先法完成壁炉膛和烟囱膛。以上从建立起受测者的知识架构再追踪其搜寻路径的方法，被用在 20 张图中做数据分析。

分析结果证明本实验中的新手在其 10 张图中，都用广度优先法。至于专家组，5 张图是用广度优先法，3 张图是用深度优先法（这三张图里分别为 90%、

drew for any object) are selected from the same drawing. At the moment when these lines are to be drawn, one of two decisions is likely to be made. The first decision (representing category one) is how far away to start the second line. All the second lines observed in this category are parallel with, and have the same length as the first lines. The second decision (category two) is to decide how long the second line should be. All the second lines in this category were perpendicular to the first lines. Theoretically, after the first line has been drawn, subjects have to scan the size of the mental image to determine how far away the second line should be located or how long the next line should be, which should take some reaction time.

These two scan decisions should provide an opportunity to observe the relationships between the reaction time and the size of an object. Two examples are given in Table 4-6. In these examples, the length of pause time matches the size of the drawn object. There are nine examples for category one found across four subjects, and seven show such a tendency. There are six examples of category two, but only three have a perfect match; another three examples indicate that scanning time is in proportion to the to-be-scanned distance. This result supports the finding of Kosslyn et al. (1978) that the time it takes to scan between two objects in an image is a linear function of their distance from each other. It may be argued that the observed scanning times in this experiment involve hand movement, but the hand movement is accomplished together with the scanning process. In other words, the movement of the pen serves as a pointer for the cognitive receptor system to provide spatial information for matching the image stored in memory, and then drawing starts.

Information Search Strategy: Search strategy has been the key research topic discussed in artificial intelligence. In cognitive science, it also has been explored by Newell & Simon (1972). In this experiment, three natural search methods had been verified from the data. The first one is the **depth-first search**. In design, it requires the completion of the frame (or outline) plus details of each object before moving on to another object. The second one is the **breadth-first search**, which means the completion of the frame (or outline) of each object first before filling in details. Methods for detecting the search strategy utilized by subjects were first to arrange all the symbols appearing in the twenty drawings by functional connections and to organize them into a hypothetical knowledge structure model, as shown in Figure 4-15; then the route of chunk retrieval for each drawing was retraced to discover the search strategies utilized by subjects. In the proposed knowledge representation, the position of "floor" is assumed to be embedded in the chunk of "hearth". But if a hearth does not appear in a drawing, or the fireplace does not have the hearth feature, then the floor is assumed to be the subchunk of the "firebox".

Figure 4-16 is a result of a search path from expert A in Figure 4-10 of a contemporary fireplace. The search pattern in this figure shows that, in most moves, a whole chunk is completely drawn before processing the next one. For example, the subject would sequentially retrieve the chunks of firebox before moving on to the next chunk of mantelpiece. Among the moves, 27 out of 30 moves were done using depth-first method. There were only three moves done by a breadth-first search——the successive moves after the floor (from move number

表 4-4 组集单元间以机能相连的总数、或然率以及组集的搜寻方式

	图	机能连接	总共连接	或然率	搜寻方式
新手 A	#1	2	7	0.29	
	#2	4	7	0.57	广度优先
	#3	6	12	0.50	广度优先
	#4	2	4	0.50	广度优先
	#5	3	4	0.75	广度优先
总分总和		17	34	0.50	广度优先
新手 B	#1	3	6	0.50	广度优先
	#2	2	5	0.40	广度优先
	#3	7	11	0.64	广度优先
	#4	3	7	0.43	广度优先
	#5	4	6	0.67	广度优先
总分总和		19	35	0.54	广度优先
专家 A	#1	7	11	0.64	广度优先
	#2	6	8	0.75	混合法
	#3	9	13	0.69	广度优先
	#4	7	10	0.70	广度优先
	#5	23	30	0.77	深度优先(90%)
总分总和		52	72	0.72	
专家 B	#1	13	16	0.81	深度优先(88%)
	#2	7	9	0.78	广度优先
	#3	11	14	0.78	深度优先(91%)
	#4	6	7	0.86	广度优先
	#5	11	15	0.73	混合法
总分总和		48	61	0.79	

88%和91%的深度优先法），两张图是混合法（表4-4）。这说明专家倾向于使用深度优先法。使用此法的原因可能是专家的知识网里有更多层次的节点是强烈地由建筑机能贯联起来的。因此当组集被激化后，回记就很快地蔓延到低下层的节点，因此储存在组集内的信息就能快速地被撷取。在这实验中，数据表明专家用了深度优先法，而且其平均反应时间也更短。特别是专家B用了这法两次，她的平均时间也是所有受测者中最短的。

实验总结

实验得出的结果似乎证明专家设计师有极强的心智影像能力。但因为实验样本少（见附记5），因之得出的结果可以被认为是启发性，而非决定性的。更多的受测者需要参与，更多的调查研究需要带动，以便进一步地充分体现这理论，并将这理论综合化。但是几个重要的发现可以归纳成下述几项：

（1）有经验的建筑师比外行的设计师拥有更大量的建筑象征符号清单储存于记忆中。专家设计师能提供更多的细部，而且能在图上表达出来。

（2）建筑师似乎以建筑机能把专业知识在记忆中组织起来。在调查两个连续成功回取组集的相互关系时，发现实验中的专家组有75%有机能关联存在。而外行组只有52%具有机能关系。

（3）专家组在由长期记忆中撷取与建筑相关影

1 to 2), after the dentil (from 24 to 25), and after the flowers (from 27 to 28). Because 90 percent of the moves followed the depth-first search method, thus this expert designer utilized the method to retrieve information.

Figure 4-17 is another example, showing the sequences in which novice B drew a fireplace for a living room in a house. The search sequences indicated that the frame of the mantelpiece was drawn first, then the frames of the mesh screen and the firebox were outlined next. Afterwards, the subject went on to complete the detail of the rest of the chunk. This pattern suggests a breadth-first search.

A hybrid search method that combined both depth-first and breadth-first also was used by the two experts. The example in Figure 4-18 shows that expert B applied the depth-first method to complete the mantelshelf and mantelpiece, then used a breadth-first search to process the firebox and chimney breast. The described method of retracing the search path through the proposed knowledge protocol was utilized in all twenty drawings for data analysis. Results show that novice designers in this experiment used a breadth-first search in all ten drawings. For the expert group, five drawings were done by breadth-first search, three by depth-first search (these three cases are 90%, 88%, and 91% of depth-first moves), and two by hybrid search (Table 4-4). This suggests that experts have the tendency to apply a depth-first strategy. The utilization of a depth-first search could occur partly because an expert's knowledge network has more levels of nodes that are strongly connected by architectural functions, so that activation spreads down the path faster and information stored within chunks is retrieved more quickly. In this experiment, data did show that experts used a depth-first search, and their average reaction time of information retrieval was, in fact, shorter. Expert B, in particular, used this method twice, and her average reaction time for retrieving images was the shortest of the four subjects.

Conclusions Obtained from the Experiments

Results obtained in the experiments appear to demonstrate superior imagery ability of expert designers, but because of the small sample size (see note 5), the results must be considered suggestive rather than conclusive. More subjects and studies are needed to further verify and expand the generalization of the theory. Yet, some major findings could be concluded in the following four statements.

(1) Experienced architects have a larger repertoire of architectural symbols stored in their memories than do novices. Expert designers can provide more details and present them in drawings.

(2) Architects seem to organize their knowledge in memory by architectural function. In studying the relationships of the links between two successively retrieved chunks by the expert group, 75% of the links were functional, whereas 52% of the novice's links are functional.

(3) Experts are faster than novices in retrieving architectural related images from long-term memory. Their ability to perceive and mentally operate images is better.

(4) A breadth-first search strategy is utilized by all subjects, but experts also use the depth-first strategy.

4.5 Mental Images and Design

The mental image study suggests that architects have a larger memory data base for architectural images than do non-architects, and are able to retrieve and process such informa-

像之速度比外行新手组要快得多。他们比外行新手更有能力在心里运作影像和察觉影像。

（4）广度优先的搜寻策略被所有受测者使用，但专家组也特别使用深度优先策略。

4.5 心智影像与设计

心智影像的调查研究显示出建筑师强于非建筑师，具有极大的建筑记忆信息库，更能迅速撷取并处理这些数据。这些储藏和回收的能力，也就是何以让一个建筑师，由浩瀚的档案库中有信心也更有余力地抽取视觉数据。丰盛的影像资源也提供给建筑师更多的选择，让他有更富裕的设计表达机会。这广大的数据库应归功于建筑机能的连接关系，才能把大量的建筑象征符号在记忆中凝聚成群。因为建筑师能理解体会每个建筑设计元素的机能要求，因而对元素和机能的联想能立刻建立起来帮助记忆。也因此，在回记时联想能提供回忆的线索帮助回记。所以联想在记忆和回记都扮演着十分重要的角色。虽然在这实验中并不能决定性地说出机能和回记的前因后果，但证据显示机能对大的记忆容量、组集间强的联系、有效的深度优先策略都有相当大的贡献。这可能就是创造力的根由。实验中观察到专家们有能力把不同影像综合，并组成一个有机能的崭新影像。这把影像组集重组，以及能操作多数组集的能力暗示着创造力的来源。

设计的另一个有趣现象是设计产品的某一部分，会被同一设计者赏识，采纳成一个影像，并用在随后的设计中，也有可能被其他的设计者应用在其他的设计中。也就是说一部分的设计会被照章采取。但原来做出这独特产品的原始独特过程，并不可能在影像被全部采用时重复泡制出现。设计师采用影像的过程，也就是这研究的主题，仅是由记忆中抽取影像并使用而已。但当实务专业和设计经验累积后，影像在记忆中也日益增加。这也解释了为何建筑师比外行人有更多的记忆数据。但创造影像的过程和采纳影像的过程不同，例如专家设计师之一（图4-6）用多重线条画出影像，这多重线条及笔触说明了其图形是弹性而非定性的，也暗示着影像是在创造之中。虽然在实验中极难发掘出受测者的思考过程，但这种"视觉思考"方法是设计师普遍采用的方法之一（Laseau，1980）。换言之，创出一个独特影像的首次过程，也可能就是一个创造设计的过程。

如前所提，一个设计的某一部分会被同一设计师或其他设计师看成是一个影像，而照章录用。这接收的方法可经过复印或借用（有时稍作修改），也

表 4-5 受测者所有的停顿期总数

	组集间 = A+B+C+D (毫秒)	组集内 = C+D (毫秒)	组集间 - 组集内 = A+B (毫秒)	A + B - 300 = A(毫秒)
新手 A	3070 (43 观察数)	846 (184 观察数)	2224	1924
新手 B	5702 (34 观察数)	978 (187 观察数)	4724	4424
专家 A	2842 (81 观察数)	668 (360 观察数)	2174	1874
专家 B	1469 (56 观察数)	525 (213 观察数)	944	644
总和	(214 观察数)	(944 观察数)		

tion faster. These schemes for storage and retrieval are what make an architect competent in the ability to draw on a vast reservoir of visual information. The abundant image resources also provide architects with more choices to enrich their design expressions. This vast database of images can be attributed to the significance of the architecturally functional relationships that hold the large repertoire of architectural symbols. Because architects can comprehend the functional requirements for each design element, associations were built up while they were memorizing the knowledge. Therefore, association provides a cue and plays an important role in the recall process. As such, association is critical for both memorizing and recalling information. Although the causal-effect relationship could not be conclusively determined in this experiment, the evidence indicates that functional linkages may contribute to larger memory capacity, stronger associations between chunks, more efficient retrieval of the depth-first search method, and, possibly, creativity. It was observed in the experiments that expert group subjects were able to synthesize and combine different images to create a new and functional one. This ability to reorganize image chunks and the phenomenon of manipulating a larger number of chunks suggest the sources of creativity.

One interesting aspect of design is that parts of a design product may be recognized and adopted as an image by the same designers in later designs or by other designers in other designs. That is, a part of the design may be adopted whole. But, the unique process that produced the unique product the first time is not necessarily used when it is adopted as a whole later. The process of adopting an image, which was the subject of this research, is simply to retrieve the image from memory and apply it. However, as design experience accumulates, the number of images also increases. This explains why expert designers have a larger memory data base than their novice counterparts. On the other hand, the process of generating an image is different from the process of adopting an image. For example, one of the expert designers, whose drawings were excluded for analysis (Figure 4-6), uses multiple strokes to sketch out his image. The multiple strokes imply uncertainty and flexibility, which also suggest creation of images. It is difficult to detect the subject's thinking process in this experiment, but this type of visual thinking is a common method used by designers (Laseau, 1980). It is possible that the unique process of generating an image the first time is also the process of generating a design.

As mentioned, a part of a design may be adopted whole (as an image) by the original designer or by other designers. The mechanisms of adopting images can be through copying or borrowing (sometimes followed by a modification), which is another way of generating designs and is done very frequently, especially by students of schools of architecture. Sometimes an image is adopted (copied) without understanding of either the process that generated it or the functionalities of the parts of the image, which creates an immature design. Expert designers, however, can quickly retrieve partial image chunks and recompose them logically into functionally meaningful new chunks (new images) because they understand the functional relationships between elements better than novices, and they have a large data bank to choose from. Based upon the observations made, it is suggested that architectural students should work on building familiarity with many design elements, and on

两个扫描时间案例　　　　　　　　　　　　　　　　表 4-6
例 A: 专家 B 图 #3 (第一类: 多远之外开始画第二根线)

内　容	停顿时间 (毫秒)	依照图大小排名
烟囱胸膛宽度	1100	1
壁炉台宽度	1000	2
石砌宽度	866	3
花瓶宽度	433	4
石块宽度	366	5

例 B: 新手 A 图 #3 (第二类: 第二根线该画多长)

内　容	停顿时间 (毫秒)	依照图大小排名
壁炉台长度	3000	1
火炉开口长度	866	2
窗户长度	483	3
挂画长度	233	4

都算是做设计的另类方法。这方法通常被许多人，尤其是建筑系学生使用。有时一个影像被采用（或复制）了，但其产生的过程并没被充分了解，或这影像里各部位的机能安排没被认识清楚就用了，也会造成不成熟的设计。但专业设计师却能快速的撷取部分影像组集，有逻辑地将其重新安排成有机能意义的新组集。因为他们比外行设计师更能了解单元间机能的关联，也有更多数据可选取。基于这些观察，本章建议建筑系学生应该努力熟悉多种设计单元，多了解单元（造型）之间的机能关系，以便建立起自己的影像库（Downing, 1989），增进自己的设计能力。

附记：

1. 机能这个词在本文中被广为运用。其定义与建筑史中所用的机能主义一词有别。它适用于所有物体的结构、空间、约定俗成及美术需求等。
2. 围棋是一种日本游戏，约三千到四千年之前发源于中国。玩法是将黑白两种圆棋子放在 361 个方格棋盘上进行游戏。
3. 口语出声法在中文中有不同翻译名称，也称放声思考法、口语思考法或边想边说法。
4. 外在变量是指自变量以外，凡是未被控制但可能影响依变量结果之因素。中文里又被称为外来变量、外生变量或干扰变量等。
5. 因为经费和浩大工作量的限制，参与的受测者人数只能受限于少量样本。但这研究希望能抛砖引玉，吸引更多未来的参与者。

understanding the functional relationships between elements (forms), so that they can build up a personal image bank (Downing, 1989) to improve their design abilities.

Note:

1. The term "function" is used broadly in this context. Its definition is not as restricted as in the term "functionalism" used in architectural history. It applies to the structural, spatial, customary, and artistic requirements of all objects.
2. A Japanese game for two, developed in China between 3000 and 4000 years ago, resembling chess or checkers, played with round black and white counters on a board marked 361 squares.
3. "Think out loud method" also called "think aloud method". It is called by three other terms in Chinese.
4. "Extraneous variable" means other than the independent variables, those factors that are un-controllable but might influence the results of the dependent variable. It is called by three other terms in Chinese.
5. Because of the budget and work load constraints, the number of subjects participated was limited to a small sample size. Hopefully, this study could call for more participation in the future.

第 5 章　设计中的知识运作

设计的过程是一系列心智运作，操控智慧去解决某些问题。在过程中，人类一方面是由知觉了解信息，撷取情报；另一方面是由意识去分析情报、处理数据，组构出一个满意的解答情况。这整个过程随时间而生，变化万千，也因变化而让过程更加错综复杂。但整个过程如能分出阶段依序作条理分析，脉络也就分明了。不同设计，思考的重点也相异，但每一设计的基本认知操作是相似的，一些运作方式、程序也是共通的。如能把设计思考的过程透明化，则设计大师的思路可以公开作参考，也能提供给学子学习的机会。而且透明的设计过程也可提出可视的清晰依据做明确的历程回溯，作为提供设计修改和评估的依据。那么这般透视设计的变化会使得未来设计受到极大的冲击。本章将讨论认知在设计中所牵涉的程序、机能运作、主要的认知组构在过程中所扮演的角色，以及对目前最主要而且通用的先进研究方法作一详细的解说。

5.1　问题解决理论

认知科学领域中最重要的学说是"**解题模式理论**"。实际上，在日常生活中解决问题的能力意义深长，而且这解题能力也是和智慧相关的重要行为之一。解决问题算是一种认知活动，而设计是一种解决多重问题的活动，所以设计活动本身也算是认知活动的一部分，称为"设计的认知"或"**设计认知**"。但什么是问题？问题是当人面对当他或她想要得到一些东西，但并不能立刻知道要采取什么样的行动去达到目的的一个处境；或是当某人试着要找方法达到一个目的之状况。于是解决问题就是一种思考的模式，牵涉高层次的认知过程。

解决问题的研究最先是由 1930 年完形心理学家做的早期实验开始，之后纽韦尔、邵和司马（1958）系统、提纲挈领地描绘出人在碰到不熟悉的事件时会如何应对。这早期研究通常是在实验室里进行的，试验一些在很短时间内即可解出的小问题，而且大量收集资料，探测寻求解答的过程。在 1950 年和 1960 年间所研究的问题大多数是针对固定结构的谜题，譬如如何把传教士和食人族以一条船摆渡过河、西洋棋的下法、河内塔和证明欧式几何定理的问题等等。这些问题也以当时能写出的计算机程序做了模拟，仔细研究了问题的解题策略。

在 1960 年到 1970 年间，研究开始移向寻找解决与大量语意象征有关的资料之解法，比方医学诊断和判读大型光谱图表数据等。那时的研究已由探讨小问题转向洞察更复杂的问题，以便深入到足够担当起解决人世间的实际问题。这方面的研究工作，以纽韦尔和司马（1972）的成果可作范例。但这些研究，依然是有良好"架构"，有清楚"目的"和

Chapter 5　Knowledge operations in design

The design process is a series of mental operations that manipulate intelligence to solve certain problems. During the process, humans, on one hand, are trying to comprehend data and fetch information through perception; and on the other hand, are trying to analyze and process information through consciousness for constructing a satisfactory solution situation. The process changes through time and becomes complicated during the sequence of changes. However, if the process could be separated into stages and analyzed sequentially, the context would become clearer. Different designs may have a different thinking focus; however, the basic cognitive operations and essential operational sequences are similar in each design. If the design thinking process could become transparent, a design master's design thinking could be made known, which would provide students with good learning opportunities. This transparent process could also provide tangible evidence that could track the design history for modification and evaluation. As such, design in the future would be impacted by this visible design process. This chapter discusses in detail the cognitive processes in design, their functional operations, major roles that cognitive mechanisms play in the design process, and the advanced and appropriate research methodologies applied in the field.

5.1　Problem solving theory

The most important theory developed in cognitive science is the **problem solving theory**. Practically, the ability to solve problems has significance for everyday's life and is one of the most important behaviors that associate with intelligence. Problem solving is a cognitive process. Design is a form of multiple problem solving activities, and thus is a part of the cognitive process termed "**design cognition**". But, what is a problem? A problem exists when a person is confronted with a situation where he or she wants to accomplish something and does not know immediately what series of actions can be performed to achieve it or how to find the means to achieve the goal. Problem solving is a form of thinking which involves the use of high level cognitive skills.

Beginning with early experiments by Gestalt psychologists in the 1930's. Newell, Shaw & Simon (1958) systematically outlined how hu-

"限制"，并依赖"特定知识"领域的问题。但也因为对人类解决问题的技巧有了相当充分的了解，这时一个新的研究领域——结合计算机科学和人工智能开始浮现，这就是"**专家系统**"。大部分的专家系统是计算机程序，具备着专家解决某些问题的智慧和所需情报。这领域在工业界和工程界的应用有相当的影响。

到20世纪80年代至20世纪90年代，研究才转移到有"复杂目的"，没有特定"目标"，而且在解题过程中其本质会变动的问题。设计即是其中之一。这时，一个新的名词"**非明确界定问题**"出现（或称弱构问题），用来区分"**明确界定问题**"（或称良构问题）。明确界定问题出自于自然科学的领域（Simon，1973），可依赖固定的条理去逐步解决问题。这些条理都有方程式或公式可互相对应。但因为解决这类问题的步骤有限，所以答案也有限；只要找到正确的方程式或解法，而且过程依序执行，则问题即可迎刃而解。至于非明确界定问题，则归属于人文社会科学领域内的问题（Reitman，1964；Newell，1969）。这类问题有极广阔的问题空间，亦即该类问题可化成许多子课题。任一子课题都可再分成另一子课题。它们没有特定的"过程程序"和"目标方向"，任何程序都可获得某种满意的但非最理想的解答。因此，不明确界定问题就有极强的创造力存在。20世纪90年代至今，解决问题方面的研究获得较少的注意和成果，这可能是因为大多数学者都认为这一领域已得到充分的了解。但在设计的学科里，已逐渐开始有研究探讨解决设计问题的现象和方法。

5.2 问题解决理论的概念

最早但重要的观念是源自"完形心理学"，说明一解决问题有固定的阶段秩序（Wallas，1926）。这些秩序包括四个阶段：

(1) 准备期：在准备阶段，要解决的问题已被确定。解题者试着了解问题，收集有用资料，而且初步尝试解决问题。

(2) 运思孕育期：孕育期发生于当解决问题的初步尝试失败时，问题即被搁置于一边，但还是不经意地在处理问题。

(3) 启发期：经过筹划孕育期之后，问题的解决答案可能会似心领神会的成熟识见，灵光一闪、豁然开朗地突然爆发显现出来。

(4) 验证核实期：有时灵光一闪、突然显现的答案不见得会是真正的答案。因此需进一步确认、核对这突然爆发出的是可行的答案。

这四个阶段的观念可用来解释有创意的解决问题之心智活动。但另外从信息处理的角度来看，问题解决的观念却是由纽韦尔和司马（1972）提出。在这一重要观念中，任何问题都包含着解题者的开始状况，称为"起始状态"，以及一个"目标状态"，即是问题已被解决的结束点。由起始状态到目标状态的整个过程可被看成是一系列的转换，产生连续的状态，并可将这一过程模式化（图5-1）。任一问题状态即代表某一特定的状态点，在该点解题者会知觉到一些事件，因此这种状态也被称做"知识状态"。集合所有解题者可能达到的状态，即称为"**问题空间**"。在空间中，由一状态转到另一状态的方法由操作单元依规则而促成。这些操作单元就是驱使达到最后目标，推动个程的原动力。在此观念中，任何解题者的问题空间都包括：对目标的领悟、察觉目前所处状态与目标的关系、得出能达到目标，以及解出问题所需的策略等。图5-1简单列举"明确界定问题"的问题空间。点代表问题状态或知识状态，箭头线代表将状态往前推动的操作单元。"明确界定问题"空间仅需少数的知识状态和操作单元即可获得解答。有些问题的目标状态单一，只存在一个解答，而有些问题例子则具有一些有限的目标状态，也容得下一些有限的解答。

mans respond when they are confronted with unfamiliar tasks. Experiments are usually studied in a laboratory setting, using problems that can be solved in short periods of time and often looking for a maximum data on the solution processes. Problem solving studied in the period between 1950's and 1960's included structured puzzle-like problems, for instance, transporting missionaries and cannibals across a river, chess moves, Tower of Hanoi, and proofs for theories in Euclidean geometry; that were also simulated by computer programs written at that time to clearly explore the problem solving strategies.

In the 1960's and 1970's, research had moved to finding methods for working with large bodies of semantic information, for instance, medical diagnosis and interpreting mass spectrogram data etc. At that time, research moved from studying simple problems to more complex ones, to obtain insight on the processes involved in solving real world problems. Some good examples could be found in the studies conducted by Newell and Simon (1972). These studies are on tasks that are well structured, with clear-cut goals and constraints. With this level of understanding of problem solving, new **expert system** research, combining computer science and artificial intelligence, emerged. Expert systems are computer programs primarily designed to carry intelligence and information documented by experts in solving problems. Its application became influential in the fields of industry and engineering.

In the 1980's and 1990's, research focused on understanding problem-solving tasks when the goals themselves were complex and sometimes ill defined, and when the very nature of the problem was successively transformed in the course of exploration. Design tasks were one area of focus. At this stage, a new term—**ill-defined (ill-structured) problem**—is used to differentiate from a **well-defined (well-structured) problem**. Well-defined problems are the problems existing in the fields of natural sciences (Simon, 1973), which could be resolved sequentially by fixed rationale. These rationales correspond to certain formulae or rules. The steps to solve these problems are limited with limited solutions. As long as the correct formula and rules are found and applied to the fixed routines, problems are resolved. Ill-defined problems are problems in humanity and social science areas (Reitman, 1964; Newell, 1969), which have large problem spaces. An ill-defined problem could be divided into sub-problems which could be further divided into another level of sub-problems. Because ill-defined problems do not have fixed problem solving sequences and goal sequence, any procedures could lead to a possible satisfactory solution but not an optimal one. Therefore, ill-defined problems have the possibility of creativity. In the 1990s and presently, less attention has been paid to problem solving research, presumably because many researchers believe that the field is sufficiently well understood. However, there are some efforts to search further to explore the phenomena and methodologies of problem solving in the field of design.

5.2 Concepts of Problem solving theory

The very early but important concept of problem solving from Gestalt psychology states that there are fixed sequences of stages (Wallas, 1926). These four stages include:

(1) Preparation: A problem is recognized, the problem solver tries to understand the problem, gather useful information and make the initial attempts to solve the problem.

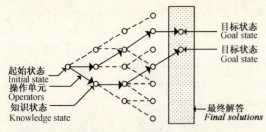

图 5-1 明确界定问题的问题空间

Figure 5-1　A problem space for a well-defined problem

完形心理学及信息处理学两个理论都有一些相似性。通常解决问题的活动涉及几个重要的工作部分：即经由注意力去选取需要考虑的议题，设立目标来考虑这些选定的议题，设计或寻找适合的行动去履行这些目标，然后由几个产生出的方案中评估选取一个满意的答案等。在这些活动课题中，前三项的界定问题方案、设定目标、规划行动等，可算是解决问题类。但最后一项评估及选取答案却应算是决策制定类。这一简略的概述通用于解决实际日常生活的问题。但在解决建筑设计问题时，又面临极大的挑战。因为建筑设计有"非明确界定问题"的本质存在。

"明确界定问题"具有清楚界定的起始状态和目标状态。"非明确界定问题"可能有模糊的起始状态、不明确的目标或不清楚的策略，以及不清晰的限制等。设计即落入弱构问题类里。设计问题的特点是因为由状况转到状况时，目标会变含糊，问题的组构会转移。建筑师面对的设计即是个很好的例子说明"非明确界定问题"类。在建筑设计开始时，设计师会有业主所要求的一般规格计划，这代表起始状态。当设计进行时，原先设定出的目标会跟着修改，也会被设计师做大幅度的调整。当设计往前推动时，当初记录在建筑草图、草模或原始简图中的早期设计概念，也会诱导产生出新标准、新可能想法或新机能要求。在整个过程中浮现出的新概念又会提供连续的回馈，并提醒设计师做额外的考虑。因此，设计过程是非常弹性、复杂的，尤其，设计师很难确定何时会是设计的终结点。

图 5-2 简示了存在于解决"非明确界定问题"中的空间元素。在空间中，可能有不少途径可获得一些满意的解答，因此解法无限。建筑设计里，建筑设计师也必须随时设立准则及限制，以便决定最后的解决方案是否可接受，以及设计过程是否可结束。也因为设计过程的复杂性，设计师一方面会策划制定问题，另一方面则构架答案。策划问题是辨认一些让设计者自己要注意的事，并且把这些要探掘的前后脉络给有条理地框架起来，所以设计师能在这框架里发掘解答 (Schon, 1983; 1988)。任一问题会在问题开始时被架构出，也会在设计课题里重复做架构 (Goel & Pirolli, 1992)。这整个问题的框架就是**"问题的结构"**。而做框架的历程称为架构问题过程。综合所有产生的部分解答途径就形成解答之结构。但问题结构是影响搜寻策略产生解答的主要因素。这些都是在设计思考过程中会发生的认知活动。

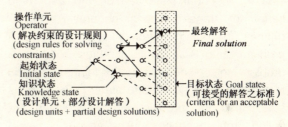

图 5-2　"非明确界定问题"的问题空间

Figure 5-2　A problem space for an ill-structured problem

5.3　表征呈现和搜寻策略

在解决问题的过程里，两个主要的因素对提高解决问题的效率有极重要的作用。第一个是表征呈现，表征呈现有助于认出可用的解题方法，以达到目标。第二个则涉及搜寻解答的策略。关于表征呈现，考虑下列问题："如何将 10 个硬币排成 5 条直排，每排有 4 个硬币？"这是传统解谜题的例子。解

(2) Incubation: The initial attempts to solve the problem do not succeed, the problem is set aside and unconsciously processed.

(3) Illumination: After the stage of incubation and some efforts done for a while, the solution to the problem may occur with a sudden burst of insight.

(4) Verification: The sudden burst of insight may not be the real solution. Verification is needed to confirm that the insight is indeed a workable solution.

These concepts can be used to explain the creative mental activities of problem solving. Other concepts have been explained by Newell & Simon (1972) in information processing approaches. In that important concept, a problem contains the initial situation of the problem solver. This situation is referred to as the initial state. The goal state is the stage at which the problem has been resolved. The process of problem solving from initial state to the goal state can be modeled as a series of transformations generating a sequence of problem states (Figure 5-1). A problem state is a particular stage in which a problem solver knows a set of things, and is referred to as a knowledge state. The various states that the problem solver achieves are called **problem spaces**. The various ways of changing one state into another are done by operators via certain rules. These operators are what drive the process forward to achieve the final goal. The problem spaces include the perception of the goal, comprehension of the present state in relation to the goal, and the possible strategies one may need in order to reach the goal and solve the problem. Figure 5-1 shows the problem spaces for a well-defined problem. Dots represent problem states or knowledge states, arrow lines represent operators that move the states ahead. The well-defined problem spaces have finite knowledge states and operators needed for solving the problem. In some cases, the goal state is unique and there is only one solution acceptable, other cases have limited final states representing limited number of solutions.

Gestalt psychology and information processing approaches have similarities. Typically, a problem solving activity involves several significant components. It is the work of choosing issues to be considered through attention, setting up goals to address the chosen issues, finding or designing suitable courses of action to implement the goals, and evaluating and choosing among alternative actions to achieve the satisfactory solution. In these series of tasks, the first three activities of fixing problem scenarios, setting goals, and plotting actions are related to the area of problem solving; whereas the last stages of evaluating and choosing solutions relate to decision making. This general description usually is true when solving real life problems. Yet, when solving an architectural design problem, it is even more challenging because the problems are ill-defined in nature.

Well-defined problems have a clearly defined initial state and goal state. Ill-defined problems may have a vague initial state, unclear goal, or unclear strategies and restrictions. Design problems fall into the ill-defined problem category. Because ambiguous goals and shifting problem formulations from stage to stage are typical characteristics of design problems, the work of architects offers a good example of what is involved in solving ill-defined problems. A designer begins with some general specifications of what is wanted by a client, which stands for the initial state. The initial goals are modified and substantially elaborated as the designer proceeds with the task. Initial design ideas, shown and recorded in drawings, models, and diagrams would suggest new criteria,

决的方法有不少，因人而异。有人会实地操作，拿10个硬币和5根杆子做实验。但实地操作不如画图方便。解谜题的窍门在于找出5根线的交互排列具有10个交集点。每个点代表一个硬币，并且是两根线的交集。解答如图5-3所示。这就是如何将问题以图形的方式呈现，以利于破解。

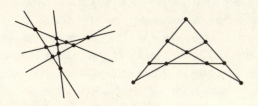

图5-3　10个硬币排列的谜题解答
Figure 5-3　Solutions for the ten coins on five rows puzzle

第二个促进解题效率的主要因素是搜寻策略，这与将起始状态逐渐转移到目标状态的操作秩序有关。在大量的可能数据中，为了提高效率，解题者通常会运用策略引导，做些选择性的数据搜寻以减少工作量。几种在研究实验室中被证实，而且发现通常会被使用的策略有下列数项：

（1）**手段-目的分析法**：这种策略，解题者会比较现存状态与目标状态的情况，找出两者差异之处，然后由记忆中寻找一些大致可能减少分歧、缩短差距的行动。

（2）**尝试-错误法**：试误法中，许多不同的理论或方法手段都会被一一探试，然后把错误的逐一剔除，一直到找到正确解答为止。此法相对来说不是很有效，大部分人都在穷途末路时使用此法。

（3）**启发诱导法**：启发诱导法是使用大拇指法则（或经验法则、或基本常识），即是基于已有的知识和过去解题累积下来的活动经验，而衍生出来的法则。这种基本法则可看成是一种将任何可想到的常识，假设为解决当前问题的最佳策略，而使用的方法。

（4）**算法**：算法是在有限数目的清晰过程中，逐一执行一套规则，以解决问题。此法是做出一些规则来测试每一个潜在的答案。本法比较适合于有明确的起始状态和目标状态的"明确界定问题"群。

（5）**假设性推理**：假设性推理是在原有问题的考虑要素环境之外所建立的设想或假说。这建立的假说不在原有设定课题的任何部分之内。

（6）**分裂而征服**：当遇到"非明确界定问题"时，一种普遍的策略是把问题分解成几个明确界定的子问题，然后逐一各个破解。

当然没有任何策略永远是最好的。有些策略也只适用于某类问题。当一个问题已完结，解题过程经验可再使用。因此，解某种问题的先前成功经验会创造出一个**心套**（或心向作用，见附记1），让解题者倾向于重复使用同样的方式解同样的问题。这种情况可能会对启发诱导式策略造成一个障碍。因为解题者会先入为主，执著于先前的成功答案，而忽略了其他简单可用解法的存在。

5.4　原案口语分析概念

大脑就像一个黑箱子，无法透明地显示出思考的过程。要了解思考过程，掌握细部都不太可能。因此要调查人类如何解问题、做决策，最重要的是收集到能探测人类行为、推理和动机的幕后资料。这些心智数据可以由面对面地询问一些特定问题而得到。最早收集心智资料的方法是1890年左右完形心理学者研究人类行为所用的内省式口语报告法（Ericsson&Simon，1984）。**内省式口语报告法**（参见第7页）是让受测者回报测验者的询问或内省回答出一些关于过去、以前已做过的行为。此法已被引用甚久。在设计中，曾有学者用录音带访问设计师研究设计过程中所用的操作单元（Darke，1979），以及解题行为的认知过程（Thomas&Carroll，1979）。在内省报告的口语形式里，受测者会报告一些推测的或自己建立的数据，反而不是回记起相关的资料。因此，此法会事过迁境后做反省，难免数据会有亡

new possibilities, and new functional requirements as the design moves ahead. Throughout the whole design process, the emerging conception provides continual feedback that reminds the designer of additional considerations that need to be taken into account. The entire process is very flexible and complicated, making it difficult for designers to finalize the goal state.

Figure 5-2 indicates the elements existing in the problem space of an ill-defined problem solving task. In this space, there could be a number of paths for achieving a number of acceptable solutions. Thus, solutions are unlimited. In architectural practice, however, a designer sets up criteria and constraints from time to time to determine if a final solution is acceptable and whether the design process should be terminated. And because of the complexity of a design problem, designers must formulate problems on one hand and frame solutions on the other. To formulate a problem is to identify the things to which a designer will attend, and to frame the context in which a designer will explore (Schon, 1983; 1988). A problem is framed at the beginning of the design and will reoccur throughout the task (Goel and Pirolli, 1992). The entire frame of the problem is the **problem structure** and the process of framing is the problem structuring process. The entire path on the generated sub-solutions is the solution structure. The problem structuring, however, is the dominating factor that influences the search strategies for creating solutions. These are the cognitive activities happened during the design thinking processes.

5.3 Representation and search strategies

Throughout the process, two major factors are important for facilitating the problem solving efforts. The first one relates to the type of representation that could be used to help identify the means available for reaching the goal state, the second one relates to the search strategies for finding solutions. In regards to representation, consider the following problem of "how to put ten coins in five straight rows with four coins in each row?" This is a traditional puzzle problem example. Different individual might use physical objects of sticks and coins to demonstrate the solutions. Yet, physical objects are not as convenient as diagrams. The key factor of the puzzle is to find the composition of five lines with ten joint points. Each joint point represents one coin and the joint point is crossed by only two lines. The solution is shown in Figure 5-3. This is an example of showing the problem with drawing representation to more easily reach the solution.

The second major factor on facilitating problem solving is the search strategies, which relate to the sequence of operations that will transform the initial state into the goal. Through large sets of possibilities, problem solvers usually apply some strategies to guide their selective search for increasing efficiency and reducing word load. Several common strategies verified and discovered through experiments in research labs are summarized below:

(1) **Means – ends analysis**: The problem solver compares the present situation at the current state with the goal, detects a difference between them, and then searches memory for actions that are likely to reduce the difference.

(2) **Trial-and-Error**: Various means or theories are tried out and faulty ones are eliminated until the correct solution is found. This method is relatively inefficient and would be used only when other solutions are not apparent.

(3) **Heuristics**: Heuristics are **rules of thumb** based on knowledge and past experience accumulated from previous problem solving activities. They are seen as a hypothesis about

羊补牢之嫌。相似于期末评图答辩时，学生会做出可能的理由辩解设计意图。为免除此困境，新的**追忆口语法**兴起了。

追忆口语法是安排受测者做一些心理实验，然后在做完实验后，立刻询问相关的心智过程。在追忆特别事件时，受测者通常会真正回记这特别的历程。此法要受测者在做完一事之后，立刻报告刚完成而且必须以事件为主导的过程数据。由于活动才刚完成，因此心智数据不会完全被忘记而可追溯。不过事后追想回忆也不十分传真，亦有编串之虑。于是**同步叙述法**就产生了。

同步叙述法就是要当事人把当时在心里进行的任何思考全部讲出来，以便获取数据的程序。艾力克森和司马（1996）证明了同步叙述报告中，受测者不会变动任何认知过程，也能外显其关注过的信息。此法是在受测者进行某事时，同时探查受测者所需带动后续步骤的关键数据（即所谓的操作单元）。但在实验情况下收到的回答对应可能与日常生活中的行为不同。不过，经过模拟真实的情况，询问经过设计的问题，并观察真正行为，所收集到的数据应该可靠到能建立起一个值得信赖的模式。最恰当的研究方法是"**原案口语分析法**"——一种出声思考的技巧。亦即把原案的思索过程一五一十地以口语经影像记录下来成为原案数据，再作个案分析。原案之意请参考附记2。原案口语分析一词在意译及字译上相得益彰，大概可以贴切地传出其韵味。由于有了原案口语数据，学者更能把人类认知过程有效地经由计算机程序做模拟。在建筑设计的研究里，一些调查结果是以此法经过"控制实验"去确认设计中所用的心智运作和表征（Eastman, 1970; Akin, 1978; Chan, 1989），由口语数据中将认知行为模式化（Akin, 1986; Chan, 1990），以及用此法分析设计活动（Foz, 1972; Goldschmid, 1991; Valkenburg & Dorst, 1998）和设计行为（Suwa, Purcell & Gero, 1998）。其他领域使用原案口语分析法的研究也包括工业设计（Dorst, 1995; Valkenburg & Dorst, 1998）、机械工程（Lloyd, Lawson & Scott, 1996; Atman, Chimka, et al., 1999）、电子工程（McNeill, Gero, et al., 1998）、软件设计（Guindon, 1990; Davies & Castell, 1992）以及其他（Goel & Pirolli, 1992）等。

5.5 原案口语数据收集、编码及分析法

原案口语分析法是个很好的收集第一手而且是原始数据的方法，去研究人类在创造设计时的认知现象。下面列举一个原案口语分析法的运用范例，详细说明口语数据的基本收集方法、实验的执行步骤，以及数据分析的程序等。这个例子是在实验室中完成的，课题是一个住宅设计，目的在于了解一个住宅设计是如何产生的。

1）课题和受测者

实验的设计课题在这实验进行时，是卡内基梅隆大学建筑系二年级的设计题目。原来的设计要求已被简化，以适合这实验目的。设计题目是一个三卧房的单栋住宅，位于校区北边的大基地上（图5-4）。设计单元包括一个客厅、一个餐厅、一个主卧和两个供儿子与女儿使用的卧室。总楼地板面积限定在2200平方英尺以内。业主是建筑透视图专业画家。两个影像单元，包括古希腊陶立克柱式和凸窗是设计要求（图5-5）。影像单元之意是指业主自行发展出的建筑形态特别要求。设计师会认辨这型，发展出一个视觉码，存在他或她的长程记忆中，以便进行这要求的影像单元设计。影像正确的平面、立面及剖面图都提供给受测者作参考（图5-5）。要求设计这影像的目的是观察设计师在何时开始处理影像部分。

在一般住宅设计中不常见的专业作坊也被列入为另一项设计要求，目的是观察设计师如何处理不熟悉的设计单元。所有设计的数据都被减低到最少，以便识别自长期记忆中所能回收的个人设计知识数

the best strategy to use in solving a problem.

(4) **Algorithms**: An algorithm is a set of rules executed within a finite number of clear procedures for solving a problem. The method is to produce sets of rules for testing every possible solution. It works best with well-defined problems with clear initial and goal states.

(5) **Hypothetical reasoning**: This method makes an assumption that was not part of the original task environment.

(6) **Divide-and-conquer**: When faced with ill-defined problems, one of the common strategies is to break the problem down into well-defined sub-problems and solve them sequentially.

No one strategy is always best. Some of the strategies only fit with certain problem types. Yet, when a problem is resolved, the experience gained from the solving processes could be used again. Therefore, previous successful experiences with certain problems might create a **mental set** (see note 1) that makes a problem solver inclined to solve similar problems in the same way. This situation might be a barrier for the heuristic strategy because one might persist with previously successful solutions even though there are easier solutions available.

5.4 Concepts of protocol analysis

The brain is a black box that does not show the thinking process transparently. It is impossible to control details of thinking and to understand its sequences. In order to study how human beings solve problems and make decisions, it is critical to document the underlying basis of human behavior, reasoning, and motivations. Information on mental activity could be collected through interviews where specific questions are asked. The earliest method called **introspective verbalization** was developed in the 1890's by Gestalt psychologists to explore human behavior (Ericsson & Simon, 1984). Introspective verbalization (see page 8) is the response to experimenter probes or retrospective answers to questions about prior behavior. This method was used for many years. Design studies included tape recorded interviews with designers, exploring the operations applied underlying design processes (Darke, 1979), and on the cognitive processes in problem-solving behavior (Thomas & Carroll, 1979). In forms of introspective reporting, subjects, instead of recalling related information, may report information that they have inferred or otherwise generated. As such, the data might be reflective after the fact. This is similar to students' behavior in that students can use whatever justifications are needed to defend their design in the final jury presentation. For avoiding such a dilemma, a new method of **retrospective verbalization** was developed.

Retrospective verbalization involves asking a subject about cognitive processes that occurred in psychological experiments conducted at an earlier point in time. In retrospective reports of specific processes, subjects retrieve the trace of the processes. The method is to ask subjects, to report just after the process has been completed, for information about the completion of the task-induced processes. Since the event was just completed, mental activity data would not yet be forgot and is retraceable. Yet, recall after the fact is not one hundred percent accurate. Thus, it evolved the need for **concurrent verbalization**.

Concurrent verbalization is the process of obtaining data while information is verbalized at the time the subject is attending to it. Ericsson & Simon (1984) demonstrated that concurrent reporting reveals the sequence of information heeded by the subject without altering the cognitive process. This method probes, concurrently with the performance of a task, for spe-

据。受测者是卡内基梅隆大学一建筑博士班学生。在参与这实验时，他有八年的设计经验，也有在建筑师事务所工作过两年的实务经验。

2) 实验程序及结果

实验从受测者读完实验说明书之后立刻开始。绘图纸和马克笔是提供的。受测者可以绘任何需要绘的图，但最后的基地图、平面图和立面图在实验结束时是必须完成的。实验无时间限制。受测者被要求从头到尾工作时必须要**放声思考**（亦即转沉默的思考成口语的思考），口语和图都会以录像带的形式记录下来。整个实验过程大约4小时（232分钟），产生不少草图和最后的平面图及立面图。图5-6中左边是一层平面，中间上方是二层平面、下方是南立面草图，右上方是剖面图、下方是完成的最后正立面图。

3) 原案口语资料的分析法

纽韦尔和司马（1972）使用的原案口语数据分析法是由"信息处理理论"的角度，经由不同的心

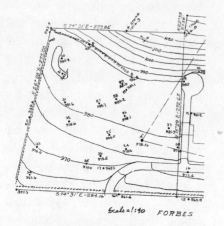

图 5-4 实验题目的基地图

Figure 5-4　Site plan of the experiment

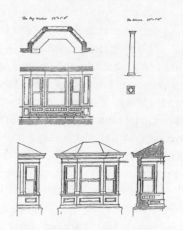

图 5-5 两个影像单元的平、立及剖面图

Figure 5-5　Plan, elevation & section of the images

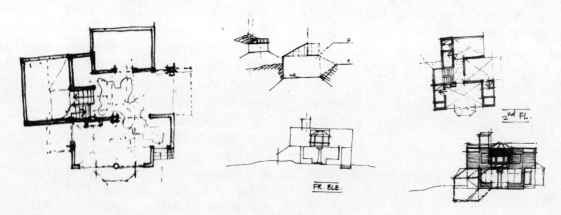

图 5-6 实验结果产生的设计图

Figure 5-6　Drawings of the design results generated from the experiment

cific information that presumably is needed to guide the succeeding behavior (i.e., **operators**). The responses in an experimental situation might differ from behavior in real life. However, through simulating the real life situation, asking a set of well designed questions and observing actual behavior, data might be reliable for achieving a realistic model of problem solving. A suitable method for gathering data is **protocol analysis**, the thinking-aloud techniques. This method is to videotape the verbalization of thoughts in real time as a case and analyze the case. For the meaning of protocol, see note 2. Through analysis of the protocol data, researchers can effectively simulate the cognitive processes through computer programs. Studies done in architectural design included controlled experiments used to identify the mental operators and representations applied in the design processes (Eastman, 1970; Akin, 1978; Chan, 1989), exploring cognitive behavior through developing cognitive models that best fit the experimental data (Akin, 1986; Chan, 1990), or using protocol analysis as a tool to analyze design activity (Foz, 1972; Goldschmid, 1991; Valkenburg & Dorst, 1998) and behavior (Suwa, Purcell & Gero, 1998). Studies in other areas of design activities included industrial design (Dorst, 1995; Valkenburg & Dorst, 1998), mechanical engineering (Lloyd, Lawson & Scott, 1996; Atman, Chimka, et al., 1999), electronic engineering (McNeill, Gero, et al., 1998), software design (Guindon, 1990; Davies & Castell, 1992), and others (Goel & Pirolli, 1992).

5.5 Methods of collecting protocol data, coding, and analysis

Protocol analysis is a good method for collecting original and first hand data to study human cognition in design creation. An example of applying the protocol analysis method is given in detail to explain the basic skills of collecting protocol data, procedural methods of executing the experiment, and techniques of analyzing data. In this laboratory experiment, the task was a housing design, and the purpose was to understand how a housing design is generated.

1) **Task and subject**

The task was used by a second year design studio in the Department of Architecture at Carnegie-Mellon University when the experiment was conducted. The original design instruction was simplified to fit the experimental purposes. The task was to design a three bedroom dwelling for a single family on a large property in the northern campus (Figure 5-4). Design units included a living room, dining room, master bedroom, and two bedrooms for a son and a daughter. The total floor area was limited to 2,200 square feet. The client was a professional architectural perspective draftsman. Two image units, a Doric column and a bay window, were required to be included in this residential design (Figure 5-5). An image unit is defined as a specific architectural form that is developed by the client. A designer was to perceive such a form and develop an image code in his or her long term memory in order to process the design task. The accurately scaled floor plan, elevation and section drawing of the image units were also provided (Figure 5-5). The purpose of having image units was to observe when and how a designer deals with image parts.

A professional workshop, which is not common in a residential design, was also required to observe how a designer processes an unfamiliar design unit. The design information was reduced to a minimum to discern the individual design knowledge that would be retrieved from memory.

理实验，系统的前后发展而出。本实验中用的方法类似于该法，但程序部分稍有调整以适合分析"非明确界定的设计问题"中对解题的思考和策略的运用。研究方法程序包括下列六个步骤：

(1) 转口语数据成原案文字表格文件

收集到的原始资料先得转成原案文字表，亦即化受测者的口语成文字表格。这些文字表格是以长于四秒钟的停顿期为界限划分出的原案叙述。一个**原案叙述**是以最基本原始的措辞方式，没经过阐释，来述说一个观察或经验。此法异于拜恩（1981）的两秒为区分界限法，因为这实验是设计过程，涉及许多图形视觉活动。一些其他的研究也说明视觉收到的数据只能在短期记忆中保存两秒（Posner, 1969），并且要花几百毫秒到两秒时间由长期记忆中搜寻相关数据（Simon, 1974）。取这些数值之上限并参考第4章中，心智影像实验所测得回收"记忆组集间"信息的944~4724毫秒之**反应时间**（表4-5），这停顿时期里由收到视觉刺激，保存视觉码于短程记忆，再由长期记忆中回收信息的全部反应时间大约是四秒钟。任何停顿时间长于四秒即表示下一个连续叙述可能是进行一个崭新的知识组集。根据这方法，本实验的原案口语存有604个文字表。表5-1列出17个原案叙述。从叙述项目1到14（4分钟时间）是了解设计课题阶段，叙述项目99到

17个原案叙述的文字表例子　　　　　　　　　　　　　　　　表5-1

#	内容
#1:	当地气候如何？<实验者: 夏季风来自东南, 冬季风来自西北。> 很好。（00:03）
#2:	看，我们有什么，有客厅、餐厅、厨房、浴厕、一、二、三个卧室和一个工作坊。（00:28）
#3:	这就是全部的区域，对吗？这房子必须要2200平方英尺吗？(00:40)<实验者: 对。>
#4:	让我看基地有多大。这(东边) 大约是110英尺，加上……(01:13)
#5:	加上大约90英尺，那大概是200英尺。(01:27)
#6:	同时这(北边) 应该稍为多于200英尺。这是120和120。那就是240, 加上多于另一边的30多英尺。因此这边总共大约是243 (计算误差)……(01:32)
#7:	嗯。这是通往住宅的马路，是在基地线里面吗？(01:47) <实验者: 对。>
#8:	同时我们有另外一条路在那(东) 边，因此这两边都是更为私密性的吗？(01:57) <实验者: 对。>
#9:	这(通道) 是15英尺宽的道路，而且……(02:23)
#10:	客厅、餐厅、厨房、浴厕，他们有没有空间要求规定客厅必须和餐厅分开, 餐厅必须和厨房分开, 或者它们可以各自为各自的部分? 他们有没有任何偏好? (02:50) <实验者: 这就由你决定。>
#11:	这一个浴室，是共用的吗? (03:14) <实验者: 你可有多于一个。>
#12:	那这三间卧室都附带浴室吗? (03:20) <实验者: 你作决定。>
#13:	不, 我问这问题的理由是因为我不确定他们的生活风格。好, 这家伙大约是38岁, 他的儿子是12岁, 女儿是9岁。他们要求分开私用浴厕吗? (03:32) <实验者: 这是中收入家庭。>
#14:	好。(04:12)
……	
#99:	另外一件很典型的事，就是把服务间，基本上是浴厕和楼梯放在同一个柱间，那就能做出分隔，让这客厅、餐厅和厨房分开，并把工作坊放外边，然后你就会把这些单元放在一块，即厨房和餐厅，或者可能把餐厅和客厅放在一块。(48:06)
#100:	我现在试着要做的东西是……(48:37)
#101:	弄好这浴厕、楼梯和厨房，因为这些基本上都是服务空间。(48:48)

The subject was a Ph. D. in Architecture student enrolled at Carnegie Mellon University. At the time of the study, he had eight years of design experience and had worked for a professional firm for approximately two years.

2) **Experimental procedure and results**

The experiment started right after the subject finished reading the task instruction. Drawing paper and marker pens were provided. The subject could draw anything he wished, but a final site plan, floor plan and facade drawing were required to be finished at the end of the experiment. There was no time limitation. The subject was asked to **think aloud** at all times while he worked, and his verbalizations and drawings were video-tape recorded. The experiment lasted for about four hours (232 minutes) generating a number of sketches plus final floor plans and elevation drawings. The drawing on the left of Figure 5-6 is the first floor plan. Second floor plan is the middle top one. The one in the middle bottom is the study drawing of the south elevation. The one on the right top is the section view, and the final facade drawing is on the bottom right.

3) **Method of protocol data analysis**

Methods of protocol data analysis used by Newell and Simon (1972) were systematically and sequentially developed from the information processing theory point of views through a number of experiments. Similar to their methods, with minor modifications to the procedure part for analyzing strategies used in solving ill-defined design problems, a methodology was developed which included five major steps.

(1) **Transferring data into protocol transcription**

The raw data was first transferred into a protocol transcription, which is a written list of the subject's verbalizations. These transcriptions were protocol statements segmented by any pause in the process greater than 4 seconds. A **protocol statement** is a statement reporting an observation or experience in the most fundamental terms without interpretation. This method differed from the 2 second pause time used by Byrne (1981) because this experiment involved a lot of visual perceptions of drawings during a design process. Other studies had shown that the visually perceived information could be held for 2 seconds in STM (Posner, 1969), and a few hundred milliseconds to a couple of seconds were needed to retrieve information from LTM (Simon, 1974). Taking the upper bounds of these numbers and referring to the "between chunk" **reaction time** of 944-4724 milliseconds found in the mental image experiments in Chapter 4 (Table 4-5 on page 86), the pause period of getting visual stimulus, holding a visual code in STM, and retrieving information from LTM was estimated in total to approximately 4 seconds of reaction time. A pause time greater than 4 seconds indicated that the successive segmented statement probably provided information about a new perceived knowledge item. According to this method, the protocol of this experiment contained 604 statements. Table 5-1 lists 17 protocol statements. Statements 1 to 14 (4 minutes) are the processes of understanding the design tasks and 99 to 101 (42 seconds) related to solving the service bay issue.

(2) **Identifying episodes**

After the protocol transcription had been completed, the transcripts were classified into episodes. An **episode** is defined as "a succinctly describable segment of behavior associated with attaining a goal" (Newell and Simon, 1972). Thus, each episode contained a unique goal that was to be achieved could be identified by: 1. the verbal information in protocol transcription; 2. tracing a series of actions which attempted to solve one design unit; 3. a par-

101（42秒钟期间）是解决服务间的设计问题。

(2) 辨认部曲

当原案口语转换成文字表格文件完成之后，文字文件即被分类为部曲形式。**部曲**定义为一描述达到一个目标行为的简短述说（Newell & Simon, 1972）。因此，一个部曲包含一个目标，目标是由：①文字表格中的口语数据认出；②追踪解决一个设计单元的连续动作分辨出；③一特定的企图解决一组设计单元来决定。表5-2是第一部曲例子，亦即将表5-1中的原案叙述项目由1到14部分定成第一部曲。这部曲是设计实验刚开始的过程，设计者的目的是收集数据，了解设计课题、要求及限制。这个阶段中发展部曲和确认目标的目的，是观察以何种认知机制设定目标，以及探讨目标如何被启动和终结。本实验中共有22个部曲被收集辨认出。

(3) 辨认知识状态

部曲被辨认之后，部曲中的知识状态是下一个被澄清的阶段。知识状态意指在某个知识阶段时，短期记忆中有某些数据存在，并且是处于活跃可用的状况。任何知识状态的改变即象征着过程是在进展中，或是正在营运某一操作单元。因此，追溯知识状态的进展基于任何在原案叙述中数据出现的变化而决定。在这研究里，辨认知识状态的目的是研究设计过程中有哪些知识出现，出现的知识和哪些设计单元有关，是哪种操作单元（或设计规则）控制了设计的进展等。表5-3列出了与设计服务间有关的四个知识状态（部分原案口语列于表5-1）。

(4) 发展问题行为图解

知识的状态或进展只提供一些片断线索。为了解设计者完成某个目标的通盘进展，"**问题行为图**

第一部曲例子	表5-2

部曲1: 收集资料情报并了解课题。

 目标 = (收集设计情报并了解课题指示)

  子目标 =收集情报：

地方气候	→(夏季风来自东南)
	(冬季风来自西北)
(设计单元)	→客厅、餐厅、厨房、3个卧室、工作坊
总楼板面积	→(2200平方英尺)
基地范围	→200英尺×243英尺
通道	→在基地产权内
两条临近道路	→两个私密边线
通道路宽	→(15英尺)

  子目标= (了解课题)：

空间需求	→客厅是否必须和餐厅分开，或餐厅必须和厨房分开，或它们可以各自为各自的部分? 有没有任何偏好? →(没有)
公用浴厕	→(可有多于一个浴厕)
3个卧室可有附带浴厕	→(中等收入家庭)

最终结果: 目标圆满达成。

Examples of 17 protocol statements from the transcriptions	Table 5–1

#1: What about local climate? <*Experimenter: Summer breeze comes from southeast. And winter breeze comes from northwest.*>OK. (00:03)

#2: See, what we have, living, dining, kitchen, bathroom, one, two, three bedrooms, a workshop. (00:28)

#3: And this is the total area, right? The house ought to be, two thousand and two hundred square feet? (00:40) <*Experimenter: Right.*>

#4: Let me see how big the site is. This (east) is about one hundred and ten feet, plus... (01:13)

#5: Plus roughly ninty feet, that's about two hundred feet. (01:27)

#6: And this (north) should a little bit more than two hundred. This is one twenty, and one twenty. That is two forty, plus thirty more feet than another side. So is two forty three approximately. (01:32)

#7: OK. This is the road that gets access to the house, inside the property? (01:47) <*Experimenter: yes*>

#8: And we have other road on that (east) side, so these two edges are more private? (01:57) <*Experimenter: yes*>

#9: This (access) is fifteen feet wide road, and...(02:23)

#10: Living room, dining, kitchen, bathroom. Do they have any space requirements that living has to be absolutely separate from the dining, anddining absolutely separate from the kitchen or they are going to be a part of the other one? Do they have any preference? (02:50)<*Experimenter: it is all up to you.*>

#11: This one bathroom here, that is common? (03:14) <*Experimenter: You can have more than one.*>

#12: And all these three bedrooms here, they have attached bathrooms? (03:20) <*Experimenter: it is up to you.*>

#13: No, the reason I am asking is that I am not sure about their life style. OK, this guy is about thirty-eight, and his son is twelve years old, and the daughter is nine years old. Now, do they require separated personal bathrooms? (03:32) <*Experimenter: This is a middle income family.*>

#14: OK. (04:12)

...

#99: One other thing is that typical, just put the service bay, basically the bathroom and staircase, in the same bay, and that might give the division, let the living, dining, and kitchen, and workshop outside, and then you want to keep these things together, kitchen and dining and then possibly, dining and living together. (48:06)

#100: The thing that I am trying to do is...to (48:37)

#101: Get the bathroom, staircase and the kitchen. Because these are basically the service spaces. (48:48)

Example of the episode one	Table 5–2

Episode 1: Information gathering and task understanding.

    Goal = (collect design information and to understand task instructions)

        Subgoal = Information gathering:

Local climate	→(Summer breeze comes from southeast)
	(Winter breeze from northwest)
(design units)	→living, dining, kitchen, 3 bedrooms, workshop
Total floor area	→(2200 square feet)
Site area	→200feet×243feet
Access road	→inside property
Two adjacent street	→ two private edges
Width of access road	→(fifteen feet)

        Subgoal = (Task understanding)

            Space requirement → should living absolutely separate from dining, and dining absolutely separate from kitchen, or they can be a part of the other, any preference→(no)

            Bathroom is common →(can have more than one)

            3 bedrooms have attached bathroom→(middle income family)

Termination: goal satisfaction.

设计一服务间的四个知识状态　　　　　　表 5-3

子目标：发展出一几何形。

目标详细：这几何形必须有些基本组合容得下一个服务间。

设计单元：浴厕及楼梯间。

　规则 1：把服务间、浴厕和楼梯间放于同一个柱间。

　规则 2：分开客厅、餐厅及厨房，并把工作坊放于外边。

　规则 3：把厨房和餐厅作一块，把餐厅和客厅作一块。

　规则 4：安顿服务间，包括浴厕、楼梯间和厨房。

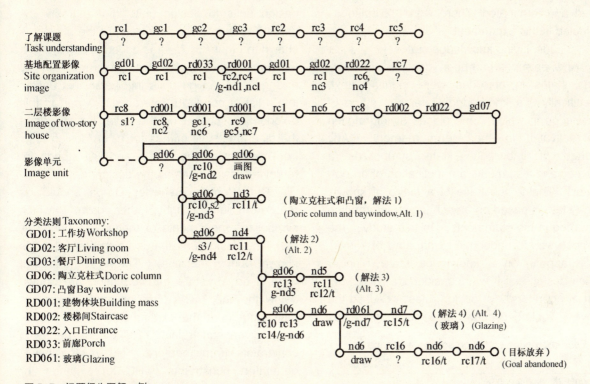

图 5-7　问题行为图解一例
Figure 5-7　An example of problem behavior graph

"解"提供了一个妥善方法。此法简洁地表示知识状态的进展。图解中节点代表知识的状态，线象征设计单元到设计单元的转移。图 5-7 是受测者解问题之一例。图解中的符码依分类法而定。这分类法是将本设计中所有设计单元依体系安排成系列条形码当做分析参考。部分分类符码列在图 5-7 左下角。GD 表示指定的设计单元，RD 是设计者自己定出的单元。

　　问题行为图解应由左到右、由上往下读。这也代表了设计的运作过程。在某知识状态时考虑的设

ticular recognizable intention under which a group of design units were to be resolved. Table 5-2 is the example of the first episode which includes statements from items 1 to 14 shown on the Table 5-1. This episode is the beginning of the design with the goal of gathering information and understanding design tasks, requirements, and constraints. The purpose of developing episodes and identifying goals were to observe the cognitive mechanism that determined goals, and to find out how goals were initiated and terminated. There were 22 episodes collected in this experiment.

(3) **Identifying knowledge states**

After episodes had been identified, knowledge states in episodes were clarified next. Specifically, a knowledge state is a stage of knowledge in which some pieces of information are activated in short term memory. Any change of knowledge state symbolizes a move, and also marks an application of an operator. Thus, the trace of a move of knowledge state is based on any changing information occurring in the statements. In this study, the purpose was to understand what kind of knowledge appears in a knowledge state, under which design unit it was considered, and what sort of operators (or design rules) caused the move. Table 5-3 shows four knowledge states

Four knowledge states on solving a service bay

Table 5-3

Subgoal: develop a geometry
Goal specification: the geometry should give some basic organization in which a service bay could bit.
Design units: bathroom and staircase.
    Rule 1: put the service bay, the bathroom and staircase in the same bay.
    Rule 2: leave living, dining, kitchen and workshop outside.
    Rule 3: keep kitchen and dining together, keep dining and living together.
    Rule 4: get the service spaces which are bathroom, staircase, and kitchen

on solving the service bay issue (partial protocol statements are listed on Table 5-1).

(4) **Problem Behavior Graph development**

The knowledge state and its move only provide fragmental information. In order to understand the whole sequential moves in achieving a goal, a **Problem Behavior Graph** (Newell and Simon, 1972) is used. The Problem Behavior Graph is a concise expression of moves of knowledge states. Nodes represent knowledge states and lines symbolize transformations from design units to design units in the graph. A partial Problem Behavior Graph of the subject is shown in Figure 5-7. It is coded according to a taxonomy which is developed hierarchically by arranging all design units involved in this design into a list of codes for reference. A part of this taxonomy is shown on the lower left corner of the diagram. GD stands for given design units from the design program, and RD stands for retrieved design units by the designer.

The Problem Behavior Graph should be read from left to right, then down, which symbolizes movements of the design processes. The design unit that is being considered is shown above the line. The operations that were used for state transformation are shown below the line. The question mark represents data input from the experimenter. GC represents given constraints. RC means retrieving constraints from memory. NC means newly generated constraints. RCXX/G stands for generating a solution by applying the rules that are associated with the constraint XX. RCXX/T stands for applying the retrieved constraint XX to test the result. Draw means drawing actions. The far left vertical line symbolizes goal stack. The purposes of constructing this Problem Behavior Graph were: 1 to observe how a goal was achieved; 2 to understand the pattern of moves; 3 to detect how search

计单元列于线上方,推进状态的操作单元则列于线下方。问号代表经由受测者询问而得到的数据输入。GC 代表给予的设计限制,RC 是受测者由记忆中回记的设计限制,NC 则是新产生的限制。RCXX/G 表示经由使用和限制 XX 相符的规律而产生的解答。RCXX/T 是使用回记出的限制 XX 来评估所产生的解答。画表示画图的动作。最左边的直线象征一叠有待完成的目标。发展这问题行为图解的目的是:①观察目标是如何达成的;②了解进展的形式;③了解搜寻方法是如何执行的。图中知识状态的阻塞回返并不代表该知识被中途放弃,却意味着目标有进一步的发展,或搜寻方法的某些效果促成知识状态的变化。

(5) 寻找不变化的结构

由观察问题行为图解中典型的过程形态,将数据输入能描述设计行为的认知模式中(Chan, 1990)查看恒定的设计过程,则一些稳定恒常,并能代表设计者一般设计行为的认知结构,应可露出一些端倪。

(6) 资料收集的可靠性

为探试资料收集的可靠性,原案口语文字表的一部分(即原案叙述项目从 49 到 86,解决希腊陶立克柱式和凸窗设计的过程)交由第三者——一位建筑师,发展出另一个"问题行为图解"做可靠性比较。在这测试中,一套详细的过程指示交给这建筑师,并要求他遵守执行的过程。虽然这第三者并没有深入了解本实验的主题,但依程序做出的结果却显示出不少和图 5-7 相似之处(图 5-8)。特别是解决陶立克柱式和凸窗设计的四个"产生—并—试误"的搜寻循环周期,都出现在两个图形里(比较图 5-7 及图 5-8)。这证明用观察知识状态变化以探寻策略的方法是恰当的。同时,采用相同方法辨认出的知识状态和得出的操作单元(即设计规则)也与图 5-7 吻合。这表示只要使用同样的方法,则知识状态和驱动设计进展的操作单元都可由数据中区别出来。因此,这发展出的原案口语分析是提供一个可

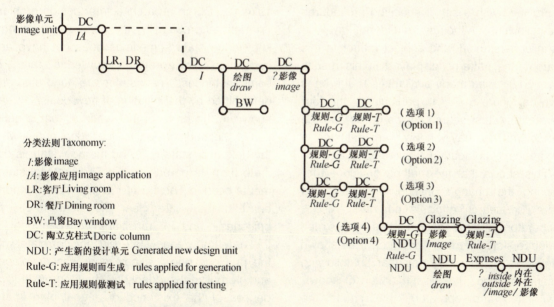

图 5-8 探测问题行为图解的可靠性和有效性

Figure 5-8　Test of the validity and reliability of the problem behavior graph

methods were implemented. The back-up of a knowledge state does not mean that the knowledge has been abandoned. Rather, it signifies the change of knowledge state that corresponds to either goal development or searching effort.

(5) **Discovering the invariant structure**

From observing the typical processing pattern exhibited in the Problem Behavior Graph, and from fitting data into a defined cognitive model describing a design behavior (Chan, 1990) to observe the consistency of the design processes, the invariant cognitive structure that represents the system's general design behavior can be detected.

(6) **Reliability of data collection**

In testing the reliability of data collection, a portion of the protocol transcription, (the process of statements from 49 to 86 on solving Doric column and bay window) was given to a third person, an architect, for the purpose of coding the Problem Behavior Graph. He was provided with a set of specified procedures to be followed. Although he did not have insight in understanding the subject matter in this research, the patterns displayed in his results (Figure 5-8) had many similarities to the ones shown on Figure 5-7. In particulars, the four generate-and-test search cycles on solving the Doric column and bay window were detected in both graphs (comparing Figures 5-7 and 5-8). This shows that the method developed for observing the change of knowledge state to explore the search effort is pertinent. Also, the operators (i.e., design rules) discerned knowledge states were in agreement, as shown in Figure 5-7. This indicates that by following the same method, the discrimination of knowledge states and operators that cause the move can be made explicit. The protocol analysis method that was developed provides an exact norm for gathering data about design processes.

4) **Results of the protocol analysis**

The collected protocol data, the knowledge base, goal plan, constraints, and heuristics applied by the designer provide a clear picture about the design methods used by this designer. These analytical results provide a hint that using such a method to study a design process will enhance our understanding on how a design is generated, and what cognitive mechanisms are involved in the generation process. In the following, results of studying design thinking processes in this experiment are explained in detail.

(1) **Goal plan**

Research in well-defined problems has indicated that any information contained in the knowledge state can be used to guide the generation of new knowledge states, so that the search through the problem space can be selective rather than random (Simon and Lea, 1974). But in an architectural design problem (ill-defined problem), the generation of a new knowledge state sometimes does not have any correlation to the previous knowledge state. In some instances, new states are generated and determined by the initiation of new goals. It has shown in the protocol that when a goal is satisfied, the subject has a clear idea about what the next goal state is. And the new state is discrete from its parent state. The protocol statements cited in Table 5-4 indicate such phenomena that there are clear distinctions between episodes; and in transitions between goals, the new knowledge state did not correlate with the previous one, a new goal was developed and verbalized suddenly. These provide evidences of the existence of a goal plan in the subject's memory and new goals are retrieved from the goal plan.

靠的收集设计过程数据之方法。

4) 原案口语分析的结果

由分析收集到的口语数据,检视设计者的目标计划、设计约束和所用启发诱导法,一个清楚的图片即浮现,说明设计师在这设计中使用的设计方法。这些分析结果也提供一个引言,说明使用此法研究设计过程会帮助人们了解设计是如何生成,并在过程中有哪些认知机制是会被涉及的。本实验中,研究设计思考过程的结果详述于下:

(1) 目标计划

在"明确界定的问题"研究中显示,存在于知识状态中的任何资料都可用来引导新状态的产生。因此在问题空间中经搜寻而产生的新状态会经过选取,而非随意生成(Simon & Lea, 1974)。但建筑设计(非明确界定的问题)里的新旧状态间有时并没有相关联系。某些例子中,新状态的产生是缘于新目标的形成而定。原案资料中就有例子说明当一个目标已达成,设计者很清楚下一个目标状态。因此,新状态和前状态是分离的。表 5-4 显示出这部曲与部曲间清楚分隔的现象。亦即在转换目的时,新的目标状态与前状态毫无瓜葛。而且新状态突然间就在口语中出现。这证明设计师的记忆中有一个设计目标计划存在。新的目标之产生由回记而生成。

(2) 达成目标的策略

本研究中发展部曲的主要意图是探讨目标如何达到。其特征是每一部曲于目标已确定时起始,而于目标已达成后终结。如果一主要目标无法被完成,则一子目标或一系列子目标会依序发展出,直到这主要目标完成为止。由研究部曲中目标的前后发展脉络,将各自的子目标分开,则设计师在一设计中

目标转移的原案叙述　　　　　　　　　　　　　　　　　　　　　　表 5-4

目标: 基地规划	#15: 目前我所能想到的就是基地。(04:49)	
目标: 二层住宅	#27: 看,我正试着要做出某些影像。一个(在第二部曲中)是基地的组成,也就是主要建筑机能的设置,如工作坊要放在那个方位,以使用作阻挡冬季风的屏障,同时在垂直方向(新的第三部曲)上,如果我要把整个方案分成多层楼板,这是许可的,对吗?还是说所有单元必得放在同一层楼里?(08:22)	
目标: 房间大小	#145: 好,可能晚一点我会再试着做工作坊。(71:05)	
	#146: 现在,我要考虑尺寸大小,首先得把这些空间考虑出大概的尺度。(71:10)	
目标: 正确的一楼布局	#225: 我这里考虑的方法是,既然这(陶立克)柱子是在一楼,那在那(陶立克柱子)上面,就可以支持这凸窗……放在柱子顶面上部,因此这凸窗就变成卧室的一部分了。(101:41)	
	#226: 现在,似乎,很可能,我要试着考虑做出一些更要正确的尺寸比例。(102:30)	
	#227: 并且看看……在基地平面上看起来会像什么。(102:49)	
目标: 正确的二楼布局	#302: ……如果我必须要考虑平面图的话。(124:58)	
	#303: 那么(放一图纸在一楼平面上)。(125:34)	
目标: 生成立面	#350: 这是平面里所可能解决、体现出的,除了(工作坊天花屋顶)之我不清楚,我必得再花一些时间解决那部分,想怎样把那部分给做出来。(137:54)	
	#351: 这正立面,就是我现在要做的部分。(138:08)	
目标: 一楼的最后平面图	#457: 好,现在我所要做的就是将所有的这些平面图再画得清楚些。(174:48)	
	#458: 所以我大可将图纸放在基地平面上,来确定草图的正确方位,或者我也可以直接画在基地图上。(175:10)	

Protocol statement of goal transformation — Table 5-4

Goal: Site organization	#15: The thing that I am trying to think of right now is in terms of the site. (04:49)
Goal: Tow-storey house	#27: See, the things that I am trying to get are some kind of image. One (previous episode) is the site organization, where the major function goes, which side the workshop should go and act as a buffer for the winter breeze, and also vertically (new episode), in terms of if I am going to divide the scheme to fit in more than one floor, that is permitted, right? Everthing has to be in one floor? (08:22)
Goal: Room size	#145: Ok, maybe later I would try again the workshop. (71:05)
	#146: Now, I would try to get into the size, first to get some rough size for each of these spaces. (71:10)
Goal: Accurate first floor layout	#225: The way I am thinking here, that you have the (Doric) column on one floor, and then on top of that (Doric column), you support the bay window...on the top level, so this bay window is a part of the bedroom. (101:41)
	#226: Now, it seems, uhm, possibly, I would like to try to get into a little more accurate scale. (102:30)
	#227: And see...how it looks on the site plan. (102:49)
Goal: Accurate second floor layout	#302: ... So if I have to draw a plan. (124:58)
	#303: then (place a paper over the first floor plan). (125:34)
Goal: Elevation generation	#350: This is how the plan is solved, worked out, except for (upper part of workshop), I don't know, I still have to spend some time in resolving that, how that place is going to work out. (137:54)
	#351: As this elevation in the front, this is what I am trying to get right now. (138:08)
Goal: First floor final plan	#457: Ok, now, what I would like to do is to draw it up a little bit more clear, all the plans. (174:48)
	#458: And then I can place the sheet on the site plan, so to give an exact location. Maybe I can draw it on the site plan. (175:10)

(2) **Strategy of accomplishing a goal**

The purpose of developing episodes in this study is to determine how a goal is completed. One characteristic of episodes is that each starts from a developed goal and ends when a goal is achieved. If a goal can not be achieved, then a subgoal or a sequence of subgoals is developed until it is completed. By studying the context of goals in episodes and by excluding their subgoals, the designer's overall goal plan used in this design can be discovered as shown on Table 5-5.

The designer's goal plan reveals the following characteristics:

1. A goal plan executes the design from the schematic level to the detailed level which has certain design abstractions associated with it.

2. A goal plan is constructed in a series of stepwise refinements. In other words, a design proceeds from developing schematic concepts to gradually adding details later.

3. It is possible that a goal plan is obtained from previously learned knowledge about processing a design task. (Quite often, a priori goal plan experience serves as a blue print and a frame of reference for a new design.)

In summary, a goal plan is a general plan that a designer must know in order to solve a particular building typology and is learned by experience. When designers encounter a totally new architectural problem that has not been solved before, a brand new method (plan) is needed. Heath (1984) indicates that such a method might be individual and personal. It is possible that a new architectural problem (design project) specific to a new building type requires a new and different method.

(3) **Control strategy**

Control strategy refers to the method of selecting goals or developing solution paths. It also reveals the way or strategy in which a designer chooses to approach the final goal. In this experiment, the subject used five strategies.

A. Scenario development

设计师在实验中使用的主要目标和子目标　　　　　表 5–5

目标所代表的目标计划	目标和子目标
1. 了解课题	1. 了解课题
2. 基地规划	2. 基地规划
3. 二层楼建筑物(发展出一工作情节)	3. 二层楼建筑物
	4. 影像单元
4. 初步空间布局	5. 初步空间布局
	6. 工作坊位置
	7. 回归初步空间布局
	8. 工作坊的造型
5. 房间大小	9. 房间大小
6. 空间生成	10. 空间生成
7. 正确的一楼布局	11. 正确的一楼布局
8. 正确的二楼布局	12. 正确的二楼布局
	13. 决定三个卧房的空间
	14. 回归正确的二楼布局
9. 立面生成	15. 立面生成
	16. 工作坊屋顶造型
	17. 壁炉
10. 一楼最后平面图	18. 一楼最后平面图
11. 二楼最后平面图	19. 二楼最后平面图
12. 最后立面图	20. 最后立面图
13. 基地发展	21. 基地发展
14 评估	22. 评估

使用的通盘目标计划即可探出（表 5-5）。

　　设计师在这实验里的目标计划有下列特征：

　　①其目标计划是先做草图阶段再移到细部层次，并且连带考虑一些抽象的设计意念；

　　②目标计划发展的款式依层次逐步精炼而成。换言之，设计是先由设计概念开始，而后逐渐加入细部；

　　③其目标计划可能由先前做过的设计学到进行设计课题的知识（通常，先前的设计目标过程经验，会替新设计提供一些参考，也提供一执行蓝图）。

　　总而言之，一目标计划是一般的设计计划。设计师在解决一特定的建筑形态时，得先有计划才能实现设计，而且计划学自经验。当设计师面临全新的建筑问题时，崭新的方法（计划）必得发展出才行。希斯（1984）就指出这种方法会是个案处理而且是个人化的。因此，新的建筑形态中遇到的新建筑问题（设计案）就得需要一个新而不同的方法解决。

　　（3）控制策略

Goals and subgoals used by the designer in this experiment — Table 5-5

Goals representing goal plan	Goals and subgoals:
1. Task understanding	1. Task understanding
2. Site organization	2. Site organization
3. Two-storey house (develop a scenario)	3. Two-storey house
	4. Image units
4. Initial space layout	5. Initial space layout
	6. Location of workshop
	7. Initial space layout resume
	8. Form of workshop
5. Room size	9. Room size
6. Space generation	10. Space generation
7. Accuarte first floor layout	11. Accuarte first floor layout
8. Accurate second floor layout	12. Accurate second floor layout
	13. Decide three bedroom's space
	14. Accurate second floor layout resume
9. Elevation generation	15. Elevation generation
	16. Workshop roof
	17. Fireplace
10. First floor final plan	18. First floor final plan
11. Second floor final plan	19. Second floor final plan
12. Final elevation	20. Final elevation
13. Site development	21. Site development
14. Evaluation	22. Evaluation

At the beginning of episode 3, the subject said: "*See, the things that I am trying to get are some kind of image. One (previous episode) is the site organization, where the major function goes, which side the workshop should go and act as a buffer for the winter breeze, and also vertically (new episode), in terms of if I am going to divide the scheme to fit in more than one floor; that is permitted, right? Everything has to be in one floor? (08:22)*"

This protocol statement gives information about goal specifications for how to work on site organization in the previous episode. It also shows that a new design guideline or a new problem structure, which was to develop a two-story house, was under development. A **design guideline** can be understood as a specific goal plan which differs from a general goal plan. The general goal plan only provides a general goal sequences for a design task, while the specific goal plan specifies unique goals and is different from task to task. Under the guidance of the specific goal plan, a design problem can be well structured and the search for the solution path is possible. It is because of the specific goal plan that the design problem can be converted into a well-structured problem. In other words, an ill-structured problem is a well-defined one, but lacking the structure required to apply powerful or algorithmic search strategies (Newell, 1969). Such a phenomenon is analogous to Akin's (1986) finding of the usage of scenarios for problem structuring.

B. Image units as important design units

This strategy indicates that when there is an image unit, it should be solved first. At the end of episode 3, when the subject worked on the two-story house image scenario, he encountered the Doric column and bay window, and treated them as important units. The subject immediately developed a new episode, which was a new subgoal for solving these image unit problems. His

控制策略是指某种选择目标或发展解决途径的方法。控制策略也可透露设计师为达到最终目标而使用的设计计谋或战术。在这实验里，设计师用过五种策略。

A. 发展情节草案法

在第三部曲中，受测者说："看，我正试着要做出某些影像。一个（在第二部曲中）是基地的组成，也就是主要建筑机能的设置，如作坊要放在哪个立面以便用作阻挡冬季风的屏障，同时在垂直方向（第三部曲）上，如果我要把整个方案分成多层楼板，这是许可的，对吗？还是说所有单元必得放在同一层楼里？（08：22）"

原案口语提供了所有目标的特定详细数据，说明第二部曲是设计基地配置。同时也显示出一个新的**设计方针**，亦即设计两层楼住宅的新问题结构正在形成。设计方针可理解为一特别的目标计划，与一般通用的计划不同。一般的计划只提供手边设计案以一般的目标次序，但特别的目标计划却说明了特殊的目标，并且因案而异。在特殊的目标计划指引下，设计问题可变成良构问题，而且找到解答的途径也较容易。也因为特别目标计划的帮助，所以设计问题的本质可转化成良构问题。换言之，弱构问题会是良构问题，只因为结构不足，所以无法利用效力大、规则系统的策略获得解答（Newell，1969）。这一现象类似于艾肯（1986）的研究发现，问题的结构可用"发展情节草案"法架构而成。

B. 影像设计单元

这策略显示只要有影像单元在，影像单元应先被解决。在第三部曲末端，当受测者注意到陶立克柱式和凸窗的要求时，立刻将此二单元认作是重要单元。受测者马上发展出一子目标新部曲，解决这些影像单元。他的解决方法是使用陶立克柱式支撑凸窗（图5-6）。

C. 错误矫正

此策略发生于当设计者发现一设计单元被误导时，他会立即将错误纠正。在第六部曲中，受测者了解要求的作坊是一专业作坊而不是业余嗜好作坊，他立刻专注其上，直到找到满意解答为止。

D. 回溯策略

此策略是当设计者面对一无法解决的问题时，他会转移目标解决其他问题，而后再回头解决原先无解的问题。这发生于当他极难解决凸窗和主卧立面开窗部分的交接处时。他的方法是转移重点，先解决凸窗屋顶的形状。后因凸窗屋顶的形状提供了交接窗和墙面的一个解决方法。方法是在墙上切块延伸分离凸窗。因此，当设计者遇到一问题，而且没有设计规律可寻时，策略是解决其他设计单元，直到相关的规律被激发可用为止。

E. **情境感知**策略和视觉测试

情境在遍布运算中是"任何信息用来描述物体的处境"（Dey，2001）。在解决问题中，情境被定义为在问题空间中描述一个情势的所有信息。因此，情境感知策略是以视觉察辨在问题空间里当前情势的任何信息，这由"**视觉测试**"完成。

设计的解答随状态演进逐渐累积而成，外显于图和模型的信息也因此跟着变化。设计师必须随时收集情报了解问题状况，这由视觉完成。在解决问题的视觉研究里，这方面曾有辨认西洋棋位置的研究（DeGroot，1966；Simon & Barenfeld，1969），也有观察河内塔谜题的研究（Simon，1975）等。视觉辨识的现象也称为"**视觉测试**"，可由"生产系统"的方法将认知机构的机能作系统的阐述，并可以计算程序明确的制定过程（Simon，1975）。"视觉测试"的"生产系统"形式，在理论上是先辨认问题的情境和解答的情境，而后决定适当的下一步执行动作。因此，这"视觉测试"就被认为是设计过程中发展策略的主要控制机制。

三个例子清楚地解释了这一策略性认知现象。第一例是在选择解答方案时辨认是否遇到关键时刻的努力。例如，设计陶立克柱式和凸窗时有四个

solution was to use a Doric column to support the bay window on top (Figure 5-6)

C. Error correction

This strategy occurred when the subject discovered that a design unit had been mistakenly interpreted, and he resolved it right away. At episode 6, when the subject realized that the workshop was a professional workshop instead of a hobby workshop, he concentrated on it until a satisfactory result achieved.

D. Back-up strategy

This strategy was applied when the subject faced a problem which could not be solved at that time. He solved other problems instead and came back to it later. This happened when he had difficulty solving the junction of the bay window and the window opening at the master bedroom facade. His method was to switch to solving the roof shape first. His intention on matching the bay window roof produced a solution for the unsolved junction part. The solution was to make a cut on the wall as an offset to separate the bay window element. Therefore, when a designer encounters a problem without having any rules available, the strategy is to resolve other design units until available rules for the earlier unit are evoked and available.

E. **Context-awareness** strategy and perceptual-test

Context, in **ubiquitous computing**, is "any information that can be used to characterize the situation of entities (Dey, 2001)." In problem solving, context is defined as the information that characterizes a situation in a problem space. Thus, context awareness strategy is to visually recognize the information that shapes the current situation in the space, which is done by **perceptual-test**.

Design solutions are accumulated from state to state, information presented in external display of models and drawings changes accordingly. A designer must gather information about the problem situation from time to time, and this is done by perception. Research on perception in problem solving has dealt with the perception of chess positions (DeGroot, 1966; Simon & Barenfeld, 1969) or solving the Tower of Hanoi puzzle (Simon, 1975). The phenomena of perception have been formulated by production systems to describe the function of its cognitive mechanisms, and are referred to as **perceptual-test** (Simon, 1975). This production system format of perception, in concept, would perceive the problem context and the solution context to determine the appropriate action to be executed next. Hence, it is regarded as the major control mechanism in the design process generating strategies.

Three examples explain this strategic phenomenon better. The first one is the attempt to recognize a critical problem situation at the time of selecting solutions. For instance, in designing the Doric column and a bay window, there were four cycles of generate-and-test creating four alternative solutions. In each cycle, the solution had been generated by one **constraint** (see note 3) and tested by another constraint. For the second generated solution, which was to locate the Doric column in the center of a room as a single interior element to support the ceiling, subject indicated two considerations: 1. such a form must also match a classical vault, which would change the character of the house and the design; 2. he was not interested in doing a historical revival, therefore, this solution was abandoned. Although the second reason suggests a personal preference, the first one indicates that the subject perceived a "**critical problem situation**" at the time when a solution was generated.

There are critical moments existing in any design problem solving processes. The **critical**

"产生—并—评试"的循环周期。每一周期的解答之产生都由一**设计约束**（见附记3）制造出，也有另一设计约束做测试。第二个解答方案是把这陶立克柱子放在房间的中心做为惟一的室内单元支撑天花，设计师说出两个考虑：①这种形必须配合古典的拱顶，但拱顶会改变这屋子和设计的特性；②他不热衷于复古。因此这解决方法就被放弃。虽然第二个理由建议了个人偏好，但第一个理由说明了设计师在一解答产生时察觉到一个"**关键问题情境**"的发生。

任何解决设计问题的过程都会有关键时刻出现。"关键性的问题情况"与改变问题结构或答案路径的可能情况有关。一个问题结构，如本章5.2所述是问题的框架，同时也意味着问题构架过程中所形成的知识表征、目标计划和设计约束等的形式结果。换言之，"关键性的问题情况"会导致问题内部结构改变或重新结构的紧急情况。在这例子中，受测者察觉到当时的解答可能会引起设计的内部造型、结构和空间特性发生潜在变化。这些变化因为有古典历史因素加入，更可能会导致知识表征的重组或目标计划的变动。因此，解答方案的选取基于查看目前的问题情况，避免任何主要的变化发生而定。这避免变化的现象也与执著于最先发展出的初步设计

概念或草案有关。罗伊（1987）指出，最早的观念会深厚影响到后续的解决问题方向。即使途中有严重问题发生，设计者还是宁愿努力地让原概念执行到底，也不愿让步采取新方向。

第二个例子是察看问题结构，确定目标秩序的努力。在这实验中找到两个例子（表5-6）。第一例当受测者看到他当时设计出的方案是一很小的建筑体块，这体块不会影响建筑物在基地上的位置。因此他决定在设计后期再发展基地设计。结果显示，基地设计确实是发生在设计后段。第二例显示受测者要目测手绘线条长度大小之后，才决定要用的绘图比例尺度。这两个例子经由察看当前情境来做设计决策，也解释了设计表征在设计过程中的重要性。

第三个例子是察看曾经走过的解答历程，以便建立原因及理由来决定生成下一解答的策略。例如，受测者曾用对称当作设计约束，这约束是在他安排凸窗和陶立克柱式时生成的。经过四个"产生—并—测试"周期满足几个设计约束之后，陶立克柱式就定在凸窗中心线的下方。设计师很满意这最后的方案，于是一个中心对称的美学原则就因此成立了。之后，这对称原则即成为另一设计约束，再次用在解决客厅设计里。设计师说："因为这（在图

情境感知决定目标秩序的例子　　　　　　　　　　　　　　　　　表5-6

例1：

#33：30英尺是……什么？这是一个在大基地里的小房！
　　　这是一个极小的房子。(11∶33)

#34：这将要出现的是这么小。(11∶46)

#35：所以不管主要的方位朝向是什么，这房子可以晚一些再放到基地里。因为围着它的是一堆空地。(11∶56)

例2：

#42：我试着要体会出最先用的尺度以了解尺度大小的观念感觉。(15∶21)

#43：所以，这大约是这么长(画一水平线)。(15∶37)

#44：我试着要体会出是否可延续使用这尺度，或者在这尺度上只做草图，然后再做进一步的细部考虑
　　　之后……才做更多的空间组织。(15∶52)

problem situation refers to the possibility of changing the problem structure or solution path. A problem structure, explained in session 5.2 as the frame of the problem, also is the format of knowledge representation, goal plan and constraint establishment, and is the result of problem structuring. In other words, the critical problem situation is the state of affair or position that will lead to a possible restructuring of the problem. In this example, the subject perceived that the solution would cause the change of the interior form, material, structure and the character of space. Such changes may possibly have led him to restructure knowledge representation or to change the goal plan, since historical elements were involved. The selection of solution was based on the perception of the problem situation and the desire to avoid having a major change of the problem structure. The reluctance of changing the problem structure might be the same phenomenon as choosing a concept or scenario developed in the early stage. Rowe (1987) indicated that initial design ideas have dominant influence on subsequent problem-solving directions. Even when severe problems are encountered, a considerable effort is made to make the initial idea work, rather than to stand back and adopt a fresh point of departure.

The second example is the attempt to perceive the problem structure and determine the goal priority sequence. Two instances were found in this experiment (Table 5-6). The first case shows that the subject had perceived the size of the building mass as a small one which would not affect its location on the site, so he decided to do the site development at a later stage. The goal of the site development did appear at a later stage in the protocol. The second case shows that the subject intended to perceive the size of a drawn line for determining the drawing scale. These attempts to perceive the context for making decisions also explains the significance of design representation in the design process.

The third example is the strategy for perceiving the solution path to create rationales for determining the next solution generation. For example, the subject used symmetry as a constraint which was developed while he was arranging the position of the bay window and the Doric column. After completing four generate-and-test cycles of satisfying several constraints, the Doric column was located on the central line underneath the bay window. The designer was satisfied with the final result of the centralized

Examples of context-awareness that determines goal sequences   Table 5-6

Example one:
#33: Thirty feet is...what? It is a tiny house on a property like this! It is a very tiny house. (11:33)
#34: And it is going to be so tiny. (11:46)
#35: Then except for the major orientation, it doesn't matter where, one could place it later on on the site. Because there is so much of land around it. (11:56)

Example two:
#42: I am trying to get roughly that initial scale and gives me some idea of scale. (15:21)
#43: So, it is approximately so much (draw a horizontal line). (15:37)
#44: I am trying to see whether I am going to work on this scale, or I could work on just schematic on this scale, and then going into more details before I make any more...spatial organization. (15:52)

上的柱子及窗）就像那（给予的柱子图例）元素一样给予震撼的印象，最起码我要试着保留这（客厅）空间，也得试着在空间里维持这对称的部署排列。(113:40)"

在这时刻，设计师察觉到**解答的情境**，也选择下一个对解答情境适合的设计约束。解答情境意指解答 B 的生成和解答 A 是有直接关连的；或解答 A 的结果直接带动 B 的生成。在这例子中，解答 A（陶立克柱式和凸窗）的结果形成选择"对称原则"来解决客厅（B）的空间排列。所有这些设计活动显示设计师事实上是非常警觉于问题空间中的全部情境，或者是问题情境，也或者是解答情境；目的是经由视觉认知来明智地掌控设计过程以便创造出一个满意的设计作品。

F. 先决模型

先决模型是存在记忆中先前做过的设计问题之解答。因为解答的过程由设计师在以前的设计案中创出经历过。因此，设计的解答会铭记在长期记忆中，极易被唤醒回记，如有必要也能立刻运用出来。在设计师的心里，存有许许多多的解答心像。这些影像经过修改也可作为未来设计的潜在解决方案之一。于是，先决模型也变成启发诱导式心智搜寻解法的来源，也是**"案例式推理"**（见附记4）这领域的研究重点。在有些例子里，当影像在设计中重复使用时，绘图的速度会越来越快，而趋自动化。如果一绘图技法达到自动化的境界，设计师不必费神专注，也无须将其口语化。在本实验中发现，两个不同楼梯的影像、车库的格局、作坊的窗型、拱顶天花和斜屋顶都很简短地被口语提到，但是被迅速画出。

(4) 设计约束

理论上，设计是应用一些设计规则而创造出的。这些规则也被称为"约束或限制"（见附记3）。相似于"物体的属性"（Reitman, 1964）、"某些参数中约制的强度"（Simon, 1970）或"设计变数"（Eastman, 1970; Akin, 1978）的概念。一设计的约束可定义为"设计一单元或一组单元时必须要满足的一些要求"。这些设计约束以知识组集的形式存在于记忆中，内含一些设计规则而形成组集。这观念就是知识表征之一。于是在设计一物体时，设计师通常会由记忆中回记一些设计约束并运用其连带的规则（见附记4）做成设计。本实验案中，由资料分析所得到的"问题行为图解"里可看出每当一设计解答产生时，最少有一个设计约束涉及，同时有一个或一组规则被口语说出。这些约束有**"全域约束"**和**"局部约束"**两种。

全域约束在数据中显示大部分（12 个中有 8 个）由第一部曲的"了解课题"发展而成（图 5-7），并且在随后的部曲中重复出现。实验里，最突出的全域约束是"气候因素"，这气候因素影响到

在口语资料中发现的全域约束　　　　　　　　　　　　　　　　表 5-7

已设定的全域约束		想出的全域约束	
气候	(客厅, 窗户, 工作坊, 卧室)	光线	(窗户开口)
	(建物体块)	私密性	(客厅, 建物体块)
总楼板面积	(建物体块)	靠近通道马路	(建物体块)
土地斜坡	(建物体块)	公共浴厕	(浴厕)
通道	(建物体块)	卧室附设浴厕	(卧室)
基地面积	(基地)	对称安置	(平面, 立面)
		房间尺寸	(房间)

column to support the bay window and a symmetric centralization aesthetic principle was created accordingly. Later, this symmetry constraint was again used to solve the living room layout. As the subject indicated:

"*Since it (Doric column and window on drawing) is going to be something as striking as an element like that (given Doric column image), at least here I am trying to keep this (living room) space and try to maintain the symmetric disposition. (113:40)*"

At this state, the subject perceived the **solution context** and selected the next constraint which best fit the solution context. Solution context is the occurrence of solution B as it relates to solution A, or the result of solution A that leads to the cause of solution B. In this instance, the result of solution A (Doric column and bay window) resulted in the selection of symmetry principle for solving the living room (solution B) spatial layout. All these design activities demonstrated the fact that the designer was very alert to the entire context in the problem space, and particularly both to the problem context and solution context, in order to intelligently control the process through perception for creating a satisfactory design product.

F. Pre-solution models

Pre-solution models are previously generated design solutions saved in memory. Because the creation processes were developed by the designer in previous design courses, the solutions would be stored in the long-term memory with strong association, retrieved easily, and applied immediately whenever it is necessary. In the mind's eye of a designer, there are numerous images associated with the solutions which could serve as potential solutions for future designs through modifications to fit the new context. Thus, pre-solutions are one of the resources for heuristic mental search.

Also, it is the concentration of the field of **case-based reasoning** (see note 4). In some cases, when images are repeatedly used in design, the drawing skills become increasingly rapid and automatic. After a skill reaches an automatic level, it requires less attention and the person who is using the skill may lose ability to describe the skill verbally. In this experiment, images of two different staircases, garage profile, workshop window patterns, vaulted ceiling, and pitch roofs were verbalized quickly and drawn rapidly.

(4) Design constraints

Theoretically, a design is generated through applications of certain design rules. These rules are referred to as constraints (see note 3) Similar to the notion of attributes of objects (Reitman, 1964), bounds on the magnitude of certain variables (Simon, 1970), or design parameters (Eastman, 1970; Akin, 1978), a design constraint is defined as certain requirements that must be fulfilled in order to design a design unit or a group of design units. These design constraints are stored in memory as knowledge chunks that have a number of design rules associated with them. This notion relates to one of the knowledge representations. While designing an object, designers recall constraints from memory and apply the associated rules to finish the design. The problem behavior graph generated from the data analysis in this experiment showed that whenever a design solution was generated or tested, there was at least one design constraint involved and a rule or a set of rules was verbalized. These constraints are both **global** and **local** in nature.

The global constraints shown in the data are most likely developed during the first episode (8 out of 12) of task-understanding (Figure 5-7) and reappear at later episodes. In this experi-

空间组织（比较抽象层次），也影响开窗位置和玻璃厚度（涉及细部层次）。设计师用过的全域约束列于表5-7。表中左栏是课题要求的，右栏是设计师自己想到的约束。这些约束具有下列特征：①通常是在口语数据中开头时即已设定；②用于一组或全部的设计单元中；③可能可以用于不同的设计课题中。

局部约束只出现于一特定的设计单元。本设计实验中有47个局部约束被用到，这些局部约束显示设计者的考虑十分广阔。举例而言，涉及专业作坊设计的全域约束有基地坡度、气候和自然采光等；至于局部约束则包括其地下室位置、访客的容易接近度、噪声和通风等。很明显地，由这作坊的例子可看出，一些通常是局部约束的也可用作全域约束。这也显示了设计者对全域约束的选择会决定思路，并对创出的设计成品有极大的影响。

(5) 设计约束中的规则

知识被分类成**陈述性知识**和**程序性知识**。陈述性知识包括了解事件的概念和事实的知识，程序性知识包括了解如何去执行事情的程序、步骤和方法的知识（Posner, 1973; Winograd, 1975; Anderson, 1980）。陈述性知识考虑知识是静态信息，而程序性知识是知道如何去力行、实现一件工作的知识。程序性知识经过一段时间的练习和实践即会变成自动化的技能。但另一方面在做某些工作时，陈述性知识也会被转移成程序性类型，绘画即是一例。当绘画技巧已被训练到一定程度后，则有必要将思潮入画，以达更高画境。

"程序性知识"曾由"生产系统"这种被广为接受的人类认知技巧之表征研究过（Newell & Simon, 1972）。"生产系统"是一套正规有条理用来代表执行一事的行动之方法；当事情的条件状况被满足，则这事必定会实现。但"生产系统"比较缺乏弹性，不易解说复杂而且连续的事件。另一研究知识的方向则来自于"**基模**"理论（Rumelhart & Ortony, 1977）。基模理论比较适合说明人类组织、综合和提取有关连续发生事件的信息之能力。这理论认为所有的知识都组成单元形成许多基模。包含在基模里的除了知识本身之外，还包括这知识如何被使用的程序情报（Rumelhart, 1980）。因此，一个基模同时包含"陈述性知识"和"程序性知识"两者，也适合用来研究设计知识。

综合考虑这些概念，所有的设计约束即可以"约束基模"来代表。下举一例说明当设计基模由记忆中提取之后，所有附带的规则如何写成"生产系统"，并如何转成计算机程序码做计算机仿真。表5-8中有三个口语陈述，这三个陈述是以建筑物体块的设计单元为主，以基地隐私为约束，并转译成"生产系统"的规则。在规则1中，事实知识或了解基地里的私密角落所在地放在规则公式的左手边。将建筑放置于基地私密角落的程序知识则放在公式的右手边。行动的一边在解决设计问题里十分紧要。因为它包含主要的领域知识，也是设计师全力依赖的设计知识。由数据分析中得知公式的右方行动部分都储存有大量的领域知识。为了找出规则1的左方事实部分，这系统必得要依序执行规则2和3，以便运作规则1的右方行动，体现解答。

这三个"生产系统"的规则可以被转成基模格式。任一约束基模都有一识别名、一变量、一组规则和该变量的变量值，如表5-9所示。表中基模A的识别名是"基地私密性"，变数是"住宅"。基地中私密角落的事实常识落在基模里规则系统的左手边。把住宅放在私密角落的程序知识则落在右手边。基模的值由运算所有规则之后还原到这基模的参数里储存。如此例，为了满足隐私的约束，基模A到C必得被逐一执行才能达到解答完成的阶段。上述整个例子简介记忆中的组集如何被激化提取运作之后生成解答的现象。表5-8和表5-9例证口语陈述

ment, the most distinct global constraint was climate. The climate factor influenced the space organization (more abstract level), and also affected the location of the window opening and the glazing size (more detail level). The global constraints used by the subject are listed in Table 5-7. On the left of the table are constraints given by the task instruction, and those on the right were retrieved by the subject. These constraints reflect the following characteristics: 1. they were mostly evoked at the first episode of the protocol data; 2. they were applicable to a group or to all design units; 3. they were able to be used in different design tasks.

The local constraints only appear for a particular design unit. Forty seven local constraints were found in this design, which also reflected the designer's diverse considerations. For instance, the global constraints used in designing the workshop were land slope, climate and natural light; whereas the local constraints were its basement location, visitor accessibility, noise and ventilation. Obviously, from this workshop example, some of the local constraints could be used globally. It also suggests that the selections of global constraints made by the designer would determine thinking, affecting the generated products significantly.

(5) **Rules in design constraints**

Knowledge has been categorized as **declarative knowledge** which comprises the facts and concepts we know, and **procedural knowledge** that comprises the knowledge of procedures and methods of how to perform (Posner, 1973; Winograd, 1975; Anderson, 1980). Declarative knowledge is concerned with knowledge as static information, whereas procedural knowledge is the knowledge of knowing how to perform a task. The procedural knowledge becomes an automatic skill after it is practiced or performed for a while. On the other hand, when performing certain tasks, the declarative information is transformed into a procedural form. Drawing is one example. After basic drawing skills are learned, it is necessary to put thoughts into drawing for achieving higher levels of expression.

Studies on procedural knowledge have been approached by production systems which are a widely accepted representation of human cognitive skills (Newell and Simon, 1972). Production systems are a formal method representing actions of performing a task, which should be executed if specified conditions are met. However, production systems do not have the flexibility to explain well complex and continuous tasks. Another approach for representing knowledge came from **schema** theory (Rumelhart and Ortony, 1977), which more closely describes the ability to organize, summarize and retrieve information about connected sequences of events. Schema theory suggests that all knowledge is grouped into units and

Global constraints found in protocol data — Table 5-7

Given global constraints		Retrieved global constraints	
Climate	(LR, window, workshop, BR)	Light	(window opening)
	(building mass)	Privacy	(LR, building mass)
Total floor area	(building mass)	Near access road	(buidling mass)
Land slope	(building mass)	Common bathroom	(bathrooms)
Access road	(building mass)	Bedroom with attached bathroom	(bedroom)
Site area	(site)	Symmetrical disposition	(floor plan, elevation)
		Room dimension	(rooms)

口语与基模规则 表 5-8

口语叙述:

#18: 现在, 似乎, 这个 (东北) 角落里, 好像更有私密性。

因为这两边 (西和南) 靠近外边马路。(06:07)

#19: 同时在这 (北) 边有一产权, 这(东) 边也另有一私人产权。(06:21)

#20: 所以, 比较恰当的是如果我把这住宅设计放在这(东北角)就更好了。(06:28)

规则表征:

规则 1: 如果有一私密角落,

则放住宅在私密角落。

规则 2: 如果 A 边是私密,

同时 B 边是私密,

而且 A 及 B 相临近,

则 A 与 B 形成的角落是私密角落。

规则 3: 如果一边缘临近一私人产业,

则此边缘就是私密边。

设计知识的基模表征 表 5-9

基模 A: < 基地 – 私密性 >(< 建筑物 >)

规则 = 如果有一 < 私密 – 角落 >,

则放建筑于 < 私密 – 角落 >。

基模 B:< 私密 – 角落 >

规则 = 如果 < 私密 – 边缘 >(<A>)是私密,

而且 < 私密 – 边缘 >(<B>)是私密,

而且 <A> 和 <B> 是相临,

则 A 与 B 形成的角落是私密角落。

基模 C:< 私密 – 边缘 >(<X>)

规则 = 如果 <X> = 在一产权旁边,

则这 <X> 是私密的。

these units are schemata. Embedded in these groups of knowledge is, in addition to the knowledge itself, information about how this knowledge is to be used (Rumelhart, 1980). Hence, a schema holds both declarative knowledge and procedural knowledge, which is suitable for studying design knowledge.

These design constraint concepts can be represented as constraint schema. The following example explains that after a constraint schema is retrieved from memory, the associated rules can be represented by production systems and converted into computer coding for simulation. There are three protocol statements in Table 5-8. These statements relate to the design of the building mass under the constraint of site privacy, and are translated into rules written in production systems as shown in the Table. In rule 1, the knowledge of fact or declarative knowledge of knowing the private corner in the site is embedded at the left hand side of the production. The procedural knowledge, which is to put the building at the private corner, is at the right hand action side. The action side is critical in design problem solving, for it contains domain specific knowledge which is heavily relied upon by the designer. The data analyses show that there was a tremendous amount of design specific knowledge embedded at the action part. In order to find the fact in rule 1, the system will sequentially instantiate rule 2 and 3. Thus, rules in schemata are applied for solution generation.

The three production rules can be converted into schemata representation, which consists of an identifier, a variable, a set of rules, and a value of the variable as shown on Table 5-9. In schema A, the identifier is site-privacy, the variable is the <Building>. The factual knowledge of the private corner in the site is embedded in the left hand side of the production. The procedural knowledge, which is to put the building at the private corner, is at the right hand side. The value of the schema is obtained from evaluating the rules in the schema, and as a result, the value is returned to the variable. As in this example, in order to satisfy the privacy constraint, a series

Protocol and rules in schema Table 5-8

Protocol statements:

#18: Now, somehow, it seems that this (northeast) corner here, seems more private.
Because these two edges (west and south) are bound by outside roads. (06:07)

#19: And there is a property on this (north) side and a private property on
this (east) side. (06:21)

#20: So, things will be better, if I place things along this (northeastern corner)
side. (06:28)

Rule representation:

Rule 1: If there is a private corner,
then put building at private corner.

Rule 2: If edge A is private
and edge B is private
and A and B are adjacent,
then the corner formed by A and B is a private corner.

Rule 3: If edge is next to a property,
then this edge is a private edge.

Lisp representation

```
(putprop East' private-property' next-to)
(putprop North' private-property' next-to)
(putprop West' road' next-to)
(putprop South' road' next-to)
(defun private-edge(edge)
  (cond((equal(get edge' next-to)' private-property)' private)))
(defun private-corner()
  (prog(corner))
    (setq corner' ((North East)(South East)(North West)(South West)))
loop
    (cond((null(caar corner))(return nil))
          ((and(equal(private-edge(caar corner))' private)
           equal(private-edge(cadar corner))' private))
          (return(car corner))))
    (setq corner(cdr corner))
    (go loop)))
(defun site-privacy(building)
  (setq building(private-corner)))
```

图 5-9　仿真设计知识的 LISP 计算机程序机能
Figure 5-9　LISP functions for simulating design knowledge

如何规则化并被基模化。每一基模也可换成 LISP（人工智能所用的计算机语言）程序以计算机仿真人类智能（图 5-9）。

(6) 实验总结

在"问题行为图解"中总共有 286 个移动的动作。这些行动可归类于表 5-10。其中"画图动作类"意味着受测者是由描旧图画出新图的过程，也算是机械化的动作。至于"空移动类"意指受测者口语说出一设计单元或一约束名称，但没动作发生。这也代表受测者试图寻找适当的信息但不成功。去除这两类动作，有 210 项移动涉及运用某些已归了类的运作。其中，运用已发现并已鉴定的设计约束之动作有 171 项。这 81.4%因使用约束而造成的移动，说明设计约束是一解决问题重要的原动力。

依理类推，设计问题的解决能力由"设计约束"、连带的规则和记忆中的"先决模型"等因素的数量多少而定。这些因素的数量越多则越能增进设计能力。基模的内容曾被用来研究专家和外行人在解物理问题上的差别（Chi，Glaser & Rees，1982）。结果也发现物理专家的基模拥有更多的程序知识，这同样解释了为何基模内的规则决定了设计能力。但设计能力也被下列两个因素决定：

of schemata, from A to C, were instantiated to generate the solution. This shows how a chunk of design knowledge is activated from memory and applied for solution generation. Tables 5-8 and 5-9 demonstrate how protocol statements can be converted into rule representation and in turn can be represented by schemata. Each schema can also be converted into LISP (used in the field of artificial intelligence) functions as shown in Figure 5-9 to simulate human intelligence.

Schemata representation of design knowledge
Table 5-9

Schema A: <Site-privacy> (<Building>)
    Rule = If there is a <Private-corner>
        Then put building at <Private-corner>.

Schema B: <Private-corner>
    Rule = If <Private-edge> (<A>) is private
        and <Private-edge> (<B>) is private
        and <A> and <B> are adjacent
        Then the corner formed by A and B is a private corner.

Schema C: <Private-edge> (<X>)
    Rule = If <X> = next-to-a-property
        Then <X> is private.

(6) Conclusions obtained from the experiment

There are 286 total moves in the Problem Behavior Graph, which are grouped by categories as shown in Table 5-10. The category of draw action means that the subject traced the old drawings and that it was a mechanical motor action. The empty move category happened when the subject verbalized either the name of the design unit or the name of the constraint, but no decision had been made. This was interpreted as the subject trying to evoke appropriate information without success. Extracting the number of moves of these two categories from the total, there were 210 moves involved with the application of classified operations. Within them, the applications of identified constraint schemata constitute 171 moves. With 81.4 percent of the moves caused by the application of constraint schemata, this suggests that design constraint is a major driving force for design problem solving.

Number of design moves by operator category
Table 5-10

Operator category	Number of moves
Data input for design unit	3
Data input for schema instantiation	18
Rule application for schema instantiation	12
Application of constraint schemata	129
Application of instantiated schema value	12
Pre-solution model	9
Draw action	57
Empty move	19
Application of unidentified schema	8
Missing data of schema application	19
Total:	286

Presumably, design problem solving ability is decided by the factors of a number of constraints, associated rules, and pre-solution models stored in memory. With larger numbers of these factors, design skills are enhanced. The contents of schemata have been used to study the differences between experts and novices in solving physics problem (Chi, Glaser and Rees, 1982). Results found that the schemata of the experts contain more procedural knowledge. This explains why rules in

操作类中设计移动次数　　　　　　　　　　　　表 5-10

操作单元类	移动次数
设计单元的资料输入	3
基模表征的资料输入	18
应用规则以启动基模	12
约束基模的应用	129
应用被启动后的基模值	12
先决模型	9
绘画动作	57
空移动	19
运用无法辨出的基模	8
基模运用时资料残缺	19
合计	286

①选择基模和规则的能力。如果在基模内的规则不足以产生解答，则必得选取其他基模或其他规则。例如，气候是个全域约束，在本实验的第一部曲中发展出来，也用在决定立面开窗的方位。口语数据中显示，受测者并没有适当的规则决定哪一方位需要开窗，或哪一立面应不装玻璃。因此，客厅和厨房的开窗位置一直都没解决。但是设计师想起另一规则，即减少窗数和开窗大小，达成另一变通的解法。

②发展新约束以测试新设计出的单元。任一新设计出的单元可能是新创物。如能把现存的基模联系上新物，或为它发展出新基模，则可帮助测试新解答作策略决定。例如，当受测者把访客和地下室位置两个约束一并考虑在作坊设计中时，他将作坊降半层楼解决了问题，也同时加入光线和噪声两个新约束来评估这解法。另一例子显示当对称的约束被用来决定设计厨房门的中心位置时，两个约束同时产生做评估：一是厨房门应该不能直接由客厅中被看到；另一是门板的开阖应该不能阻碍住客，于是他作了一些修改。

总而言之，这研究证实了设计过程中有目标计划的存在，也有不同的认知策略之应用。同时也解释了"情境认知"如何掌握系统的进行。最重要的现象是约束基模中所存的规则知识是提供解决问题，并评估答案的资源。因此，约束基模应该看成是解决设计问题的一项工具；设计师在设计中也非常依赖过去的设计经验和目前察觉到的视觉数据。这些发现不但一方面详细解释了设计过程的本质，另一方面也提供了一个基本的观念架构，让设计学子了解什么是设计科学，也知道该如何发展自己的设计能力。

5.6 原案口语分析的运用

原案口语分析法在认知心理学的领域中，已有一段时日被用来研究人类解决问题并探讨认知活动。许多研究设计过程的学者也都用口语分析法，

schemata determine design ability. Design ability, however, also is determined by the following two factors:

1. The ability to select rules and constraint schemata. If the rules in a schema are insufficient for solution generation, then other rules or schemata must be selected or found. For example, climate was a global constraint which had been instantiated early at the first episode and was used for deciding the window location on surfaces. As shown in the protocol, the subject did not have appropriate rules for deciding which orientation should have window openings and which surface should avoid having glazing. Hence, the problems of the window location of the living room and the kitchen were not resolved. However, the subject was able to retrieve another rule, which was to reduce the window and the glazing size, and an alternative solution was reached.

2. The ability to develop new constraints for the test of a newly generated design unit. A newly generated design unit may not be a familiar one. Therefore, associating existing schemata to it, or developing a new schema for it, would help to test solution for decision making. For instance, when the subject combined the constraints of having visitors with the basement location for the workshop, he generated a half level sunken workshop and was able to develop light and noise constraints for the new design unit. The second example showed that after the symmetrical disposition constraint had been applied to create a centralized kitchen door, two constraints were applied: that the kitchen door should not be visually accessible from the living room, and the position of swing of the door should not disturb the user. Therefore, he made some modifications.

In conclusion, this study confirms the existence of goal plan, the utilities of different cognitive strategies in design. It also explains how context awareness controls the progress of the system. The most important phenomenon is that the knowledge contained in constraint schemata provides resources for solution generation and testing. Therefore, the constraint schemata can be seen as a tool for design problem solving. Designers rely on past experience and current visual information. Possession of skills in organizing and applying schemata determines a designer's design ability. These findings not only describe the nature of the design process in detail, but also define a basic framework for students to understand the science of design and to develop their own design ability.

5.6 Applications of protocol analysis

The methods of protocol analysis have been used to study human problem solving and explore cognitive activities in the field of cognitive psychology for a long time. Many studies in design processes also have used the protocol analysis technique that generates many valuable study results. For instance, studies carried out by scholars in architecture, industrial design, mechanical engineering, electronic engineering, software design, and educational research have enhanced the understanding of design thinking processes. These studies demonstrate that design cognition is a domain independent phenomenon. In September 1994, a workshop on "analyzing design activity" was held at the Delft University of Technology, Netherlands, to study the virtues of protocol analysis, concentrating on the data analysis. In that workshop, videotapes and transcribed protocols of a number of experiments (Dorst, 1995; see note 5) were sent to a group of researchers to perform the analysis in any form they chose.

做出了许多甚有价值的研究成果。例如建筑设计、工业设计、机械设计、资讯工程、计算机软件设计和教育界学者所做的研究，都增进对设计过程的了解。这些研究也很信服地证明设计认知是一个独立范畴的现象。1994年9月一个"分析设计活动"的工作坊在荷兰代尔夫特理工大学成立。研究口语分析的实质尤其专注于资料分析部分。在该工作坊，录制好的设计口语实验录像带（Dorst，1995；见附记5）和转化好的口语陈述数据被拷贝成套分送给一组学者，要学者们以他们自选的形式各做分析工作。

工作坊所做出的结果于1995年在"设计研究季刊"中专辑出版。在这些论文报告中，即使化口语为陈述的运作法，及转移成部曲码的方法程序并没有被清楚详细地报告过，例如个人设计组的口语长度是1~8分钟，分成28个单元，但团体组的口语则又长自1~9分钟分成45个单元（Goldschmidt，1995）。可是这组学者在分析同一套给予的资料时也产生许多有趣的发现。这些不同发现，例如从学习团队设计过程的社交活动（Cross & Cross，1995），插曲性知识（Visser，1995）用在设计活动中，到口语及视觉码用于设计过程中的现象（Akin & Lin，1995）等，相当多姿多彩。这显示口语数据是一非常丰富的数据库，也有潜力提供不同的研究探索题材。事实上，一套资料能产生20种不同的研究论文报告，也清楚地证明口语分析是一种以观察经验为主，研究设计活动的精确严密工具。

原案口语分析被研究社区所承认的是这研究必须得付出时间和精力来完成的昂贵费用。因此要让多位受测者参与，做多重实验，以观察一个相同设计课题中共通的设计形态是高代价的（但极有价值）。遑论如果要收集多套资料，以证实在代尔夫特工作坊这些有深度报告作出的结论之正确性，则是多么的奢侈。虽然在统计分析中，一个样本（数据套）不够充分作出一结论，或作出一强有力的统计叙述。但持平而论，任一单个实验是可以看成是领头的试点研究，也可随后被其他学者重复，并共同验证所发现成果的可靠性。依此目的，在网际网络上值得设立一个公众的原案口语数据库，让研究小区的学者都可一起分享资源，共同探讨设计思考过程里的现象。

5.7　口语分析的优点及缺点

原案口语分析的方法有些疵瑕，也有一些争论批评此法的极限。但这些疵瑕和极限如下述几类，可以小心地辅助以加强此法。

口语资料空白点：口语数据中可能会出现没动作发生的空白点，也无法查探出在这些空白差距中的心智过程。但是实验者可以警觉的控制情况，并且持续地在任何过长的**等待时间**中不断地提醒受测者维持放声思考的状况。实验者也可同时用"追忆口语法"的技巧，事后询问受测者以便获得相关的数据补填空洞的缺失（Ericsson & Simon，1980；1984）。

无法口语化的认知行动：思考的速度快过运动动作。在某些情况下，当动作正在进行时相随的思考过程无法立即被映现。当许多绘图动作出现时，这事尤其成真（Akin，1979）。相似的情况也在本实验中发生，每当绘图动作出现，受测者会念出将要画的图名，而后一段短时间内即专注于完成绘图的动作。这有绘图动作而无口语的现象可由三个观念解释。第一，受测者正在提取心像中的视觉码，并没有足够时间，为绘图而必须解出视觉码的快速认知活动口语化（参见章节4.4）。有时这些动作是自动化的技巧也没有必要口语化。第二，受测者不熟悉于放声思考法，而说不出心里想法，但是实验前的小型练习准备会有帮助。这结果可能会有较佳的口语数据，也可能会将运动动作的速度迟慢下来。第三，如果受

Results of the workshop were published on 1995 in a special edition of The Journal of Design Studies. In the reports, even though the algorithms for transcribing protocol coding and the procedural methods of episoding the codes were not reported clearly, the protocol of the individual designer was divided into 28 units by subject matter ranging in length from one to eight minutes, and the protocol of the team was divided into 45 units ranging in length from one to nine minutes (Goldschmidt, 1995). Results generated by the group analyzing the given sets of protocol data yielded many interesting findings. These diversified findings vary from the study of social activity in a group design process (Cross & Cross, 1995) and episodic knowledge used in design activity (Visser, 1995) to verbal and visual coding applied in the design process (Akin & Lin, 1995), among others. This shows that verbal data are a rich data mine which has potential for various explorations. In fact, a single set of data generated 20 different papers, which demonstrates clearly that protocol analysis is a rigorous research tool for empirical study of design activity.

As recognized by the research society, protocol analysis is expensive in that it requires time and energy to complete. However, it is costly (but valuable) to conduct multiple experiments on many subjects to observe common generic patterns of the same design task. It is also a luxury to collect more than one set of data to verify the conclusions generated by the thoughtful papers completed in the Delft Workshop. Yet, in statistical analysis, a sample size (data set) of one is not sufficient to draw conclusions nor to yield sound statistical plots. It is fair, however, to treat an initial experiment as a pilot study which can be repeated by other scholars to further test the validity of findings. For this purpose, it is worthwhile to set up a public protocol data bank, accessible through the Internet, to be shared by the research community on the study of the phenomenon of design thinking processes.

5.7 Advantages and disadvantages of protocol analysis

There are several deficiencies in using protocol analysis, and many arguments criticize the limitations of the method. These limitations and deficiencies, categorized by groups listed below, can be strengthened for the use of the method.

Gaps in protocol data: There could be empty spots with no action occurring in protocol data, and there is no way to detect the mental process occurring at these gaps. However, the experimenter should be alert to control the situation and constantly remind subjects to keep thinking aloud of any long **latency** occurring in the experiments. The experimenter should also use the technique of retrospection to fill in the information occurring in these gaps (Ericsson. & Simon, 1980; 1984).

Cognitive actions cannot be verbalized: Thought processes are faster than motor actions. In some cases, the thought process cannot be reflected immediately while a motor action is in process. This is true especially when many drawing actions appear (Akin, 1979). Similar occasions did occur in this experiment that, while there were drawing actions, the subject mentioned the name of the to-be-drawn object and then concentrated on drawing for a short time. This drawing with no verbalization phenomenon could be explained by three notions. First, the subject was also retrieving visual codes of the mental image, and these speedy cognitive activities (see Chapter 4) on decoding visual codes for drawing do not have

测者是在绘草图的图案思考状态下，而非是在机械化绘图时，则将思路逻辑随画图动作口语化是应该的也是必要的。

超过负担的认知：有评论指出，做口语需要的心智功能率有时会妨碍而且影响到受测者执行课题的能力。这对有动态本质的课题或执行时的时间因素是重要的课题而言是正确的。一个能化解心智能力互相比竞的方法是将受测者执行课题的过程先录像，然后受测者再补述做课题时的思想和所做的行为。在这过程中，影像可以减速或停止，以提供充裕时间描述细部。

对设计课题的误解：伊士曼（1969）提及解题者对课题的提示通常会犯错误。但测者和受测者可以在实验开始前互相把目标重新解说，把过程演练一次以降低犯错的几率。

设计课题的复杂性：实验课题的复杂性会影响要观察的主题。例如，简单的课题在本质上倾向于良构而且问题空间较小，因此容易让测者控制实验以利观察。但这种实验课题却会限制住观察搜寻策略的机会，这种机会有时在复杂的问题中才会发生。

真实的设计案：大部分的口语分析研究都是一次专注在一个设计案，而且是在实验室中经过一定的过程控制而完成。因此，口语原案分析是很特殊的研究技巧，但也无法捕捉在职业设计环境里真实案件的广大真实性去研究团队设计中：（1）职业化经营设计的思考方式；（2）真正世间问题的专业解决方法；（3）详细的团队合作效果。但是，如能把不同的追忆口语叙述、询问和同步叙述口语三种技巧同在一真实设计案的整个过程中用到，则在某一特定时间里用某一技巧研究某一特别现象和重点，会得到更多的探讨结果。

5.8 收集思考数据的其他方法

放声思考的原案数据较能接近一些心智过程。这些过程在事件过后是无法充分被回记的，即使能回记，但是在解述时也无法减少一些偏见或误差。因此它还是一种可靠的研究方法，也被用在许多专业领域。其他研究知识和思考的方法还包括采访面谈法和观察法。

采访面谈有组构性、半组构性和非组构性等三种方法用在实地或实验室的环境中。组构性采访有完全准备好而且充分组构的问题。受测者被要求以固定的次序回答问题。实验者在这种情况下只能得到他或她所特定要求的数据。半组构性采访综合一套充分组构的问卷并佐以自由的问题，通常受测者是会给予一套问题，而后实验者在必要时再询问其他需要澄清或补足细节的问题。这种方法相当普及，因为这方法会带领问题走向一个特定的方向而不偏失。非组构性问题则无事先定义好的问卷，实验者可以自由地询问任何适宜的问题去探讨课题。这方法是开创新研究的好方法。

另一产生原案口语数据的方法是观察法。亦即观察受测者执行课业的活动行为，对他们的表现做笔记、录像，并且综合追忆口语法合成为一个收集数据的系统方法。例如为了捕抓一个广阔的印象，而用仔细观察法对工业专业做研究的例子令人印象深刻（Frankenberger & Badke-Schaub, 1998）。

结合所有这些技巧收集资料研究某一专题，会是好的研究方法。例如有一研究试着寻找在设计期间，建筑师会不会在心里对使用人或住户建立一心像或某种期望，于是采访了五位设计公共住宅的建筑师，并且录下访问过程。然后观察所有画出的图和写出的文件，要求设计师回忆他们的设计过程，并以追忆口语和内省口语法收集需要的分析数据。最后分析采访的结果显示建筑师会用一些简单的设计目标达到一初草图，这第一草图也会进一步的形成一视觉心像。这视觉心像就形成了一设计解答方案。因此这些简单的设计目标和第一设计方案被认

enough time for verbalization. Sometimes, these actions are automatic skills which are not necessary to be verbalized. Second, the subject was not familiar with the think aloud technique, which could be improved by a warm up exercise before the experiment. The results could have a better verbalization, but, the motor actions would be slowed down. Third, if the subject was working on sketch mode on graphic thinking instead of mechanical drafting, then verbalization of the rationale should necessary come together with drawing actions.

Cognitive overload: It is argued that mental effort required to provide verbalization would interrupt and affect subjects' ability on performance of the task. This may be true for the tasks that are dynamic in nature and for the tasks that the time needed for execution is critical. One method to resolve this mental competition is to video record the task behavior first and request commentary on what the subjects were thinking and doing. During the process, the video could be slowed down or paused to allow spare time for detailed explanations.

Misinterpretation of design task: Eastman (1969) reported that the problem solver's understanding of task instruction often was erroneous. The problem solver and the experimenter should interpret the goals and elaborate the processes together before the experiment proceeds, to reduce the possibility of making mistakes.

Complication of design task: The extent of the complication of the experimental tasks will influence the observed subject matter. For example, simple tasks tend to be well-defined in nature and have less degrees of freedom of the problem space, allowing the experimenter to control the experiment for observation. These tasks, however, would also limit the opportunity for observing the possible search strategies which may only occur in complicated problems.

Real design project: Most studies on protocol analysis focus on one project at a time in a laboratory environment under controlled conditions. It is a very specific research technique, which is not able to be replicated in architectural offices to study: (1) the professional management of group design thinking; (2) the real world problem solving in the profession; (3) team work efforts in detail. However, certain topics and phenomena could be explored by combining techniques of retrospection verbalization, interviewing, and concurrent verbalization used at the specific time during the entire design processes of a real world project.

5.8 Other methods used for data collections on thinking

Think aloud protocol can access cognitive processes that can not be fully recalled without bias and distortion if explained after the task has been completed. Thus, it is a reliable research method used in many fields. Other techniques on studying knowledge and thinking include interviews and observations.

Interviews have three formats of structured, semi-structured, and unstructured interviews used on site or in laboratory environments. The structured interview has well prepared and well-established questions. Subjects are asked to provide answers in a fixed order. Experimenters can only obtain information on what is asked. The semi-structured interview combines a highly structured agenda with free questions. Usually, subjects are given a set of questions and experimenters ask supplementary questions to clarify points and add detail if necessary. This method is popular as it helps to guide the questions in a certain direction. The unstructured interview has no pre-defined ques-

为是设计的主要生成器（Darke，1979）。这旧例子说明结合不同方法探讨设计思考所能做出的丰盛成果，但这方向的研究必须进一步走向研究创造性的思考和创造力。

附记：

1. 心套或称心向作用，是指在学习之前已完成心理上的一套某种准备。
2. 原案意指文件的原稿，也是一个条约、协议或相似文件于批准承认之前的初稿。在计算机科学中，它意味着确定数据项的格式，以及数据传输的规律。中文以"原案"称呼十分贴切。
3. 设计约束（或设计限制）也可看成是一设计目标，必须满足并实现这目标设计才被认为是成功的。
4. 案例式推理方法：是撷取过去的、相似的处理类似案例状况及经验作为解决新的问题之参考。
5. 一位单人设计师和一组三人设计团队共同设计一束扎装置，两小时的设计时间都录制成录像带。

tion set. Experimenters are free to ask any appropriate question for exploring a topic. This method could be a good way to initiate a study.

Another technique for generating protocols is observation. Simply observing subjects performing activities, making notes on their performance, videotaping the processes, and combining retrospective reporting; it would be a systematic method for data collection. An impressive example of this technique was showing detailed observation of industrial practices in an attempt to capture a broader perspective of the industry (Frankenberger & Badke-Schaub, 1998).

A combination of all these techniques could be a good method for data collection on a particular subject matter. For instance, in a study intended to find whether during the design period the architect was thinking of an image or expectations about users, the method used was interviewing the architects of five housing projects and tape recording the interviews. Data was collected on observations of sketched and written outputs and asking designers to recall their own processes, retrospection and introspection. The analysis indicated the architects used a few simple objectives to reach an initial concept. This initial concept further generated a visual image which shaped the design solution. The initial concept and the objective are the p rimary generator of design (Darke, 1979). This well-cited example shows the fruitful results of combining different methods to explore design thinking. Yet, research directions need to move further in the exploration of creative thinking and creativity.

Note:

1. **A mental set** has certain psychological preparations that are in place before learning starts.
2. **Protocol** means the draft of a document and is the first copy of a treaty or other such document before its ratification. In computer science, it means rules determining the format and transmission of data items.
3. **Design constraint** also can be seen as a design goal that must be met in order for the design to be considered successful.
4. **Case-based reasoning** means to apply previous experiences and situations on handling similar cases as references for solving new problems.
5. An individual designer and a three-person team of designers worked on designing a fastening device for two hours were videotaped.

第6章 科技对认知的影响

认知是人类天赋能力的一部分，它赋予我们能力去认识这世界，了解这宇宙，并对应这环境。思考能力是认知的一部分，也因为有认知思考能力，它让人类创造文明并创造科技。文明和科技也同样影响人类认知和思考，而更进一步地反馈促进文明。在这例子中，文明被看成是一被共同分享并会影响知觉和行为的价值、信仰和态度体系。由于科技文化和认知之间密切且奥妙的互动性，因此学者为着寻找改进思考、认知、文明和科技的途径，而专注于研究驱动思考的原动力，并了解认知的基本现象。目的是找出方法改进思考、认知、文化和科技。从1910年认知心理学的起源直到目前，这领域中的理论发展经历过不少变化，但基本信念是：研究都永远建立于前人的研究成果之上，理论则由过去实验的成果累积发展而成。这一章，作为本书的完结篇，也试着综合这领域中所做出的成果，解释目前科技对研究导向的影响，也叙述未来可能的理论走向。

6.1 认知科学与人工智能

在研究人类智慧里，尤其是解决问题、推理、学习、语言、感知和决策制定等，最主要的贡献是"人脑像计算机"的隐喻。最根本的影响来自电子计算机的发明。电子计算机带来以符号表征仿真人心的优势。本着这隐喻，有名的**图灵测试**就被发展出来。这理论是测试机器参与拟似人类对话的能力，并让测试者在隔壁房间评定两个对话者哪一个是人，哪一个是机器。结论是计算机可做和人脑一样的课题。根据图灵（1950）的研究，电子计算机执行的观念可解释为"这些机器是企图实现任何人脑所能做的事情"。自那时开始，几个领域中的学者即开始根据复杂的表征和电算程序发展出一些关于心智的理论，于是人工智能开始成形。

人工智能是计算机科学的一个分支，研究如何赋予计算机一些具有人类智能的能力，并创出一些电算程序让计算机执行一些灵长类所能做的课题或动作。人工智能被许多学科应用到。在语言方面，运用语法发展出程序与人做对话，或做语音辨认。在计算机游戏中，人工智能被用来创造出由计算机控制的角色或物体的智能，因此一个角色的动作能和游戏中其他角色的动作相对应。基本上，人工智能有两个目的：一是用计算机的能力去增强人类的思考能力；另一目的是用计算机去了解人类如何思考。随着这些观点，信息处理理论"假设人脑是信息处理单元的理论"即被开发出来，并用于认知心理学中。也因这理论的形成和计算机的影响，把认知科学由认知心理学中萌发而成。这显示电子计算机的科技之发明极大影响了一理论的发展和学科的成立。

Chapter 6   Impacts of technology to cognition

Cognition is a human function, which allows us to recognize the world, understand the universe, and respond to the environment. Thinking ability, a part of cognition, allows us to shape culture and create technology. Culture and technology also affect human cognition and thinking and move culture forward. In this instance, culture is seen as a shared system of values, beliefs, and attitudes that influence perception and behavior. Because there are subtle interactions between culture/technology, and cognitive thinking, scholars have focused on studying the fundamental forces that drive thinking, to understand the basic phenomena of cognition. The purpose is to find ways to improve thinking, cognition, culture and technology. Although there have been theoretical changes in studying human intelligence since the formation of cognitive psychology in 1910, fundamental beliefs researched in the field will always build from previous studies, as theories accumulate from past experimental findings. This chapter, as a conclusion of the book, intends to summarize the works done in the field, explain the impacts from technologies that affect research orientations, and elaborate the possible future trend of theoretical movements.

6.1   Cognitive science and artificial intelligence

In the study of human intelligence, such as problem solving, reasoning, learning, languages, perceiving, and decision-making, the major contribution is the metaphor of brain-as-computer. Studies were originally influenced by a generation of digital computers, which provided advantages of symbol representation to simulate the human mind. Based on this metaphor, famous **Turing Test** was developed to test a machine's capability to participate in human-like conversations and allow the interrogator in another room to judge which response was human and which was machine. This resulted in the computer doing the same tasks as humans. According to Turing (1950), the idea behind digital computers is that machines can carry out any operation which can be done by human minds. After this discovery, researchers in several fields began to develop theories of mind, based on complex representations and computational procedures, and artificial intelligence came into play.

人类认知中另一重要的单元是人类解决问题的能力。这方面的研究探讨人类如何使用逻辑推理和知识来解决一些特定的问题。结合认知心理学和人工智能，另一新领域被称为**专家系统**也开始形成。专家系统又称为知识基础系统，通常是一套计算机程序，由一套知识规则组成，分析一套给予的情报去解决一特定的问题，并提供一系列动作来解决这问题。最早的例子是1965年一套演绎推理有机形态的分子结构软件。在1974年，另一以"实际规则"为构架做医学诊断的软件（MYCIN）被开发出，也被认为是这领域的典范之作。由此之后，专家系统的观念和方法即被广泛用于商业上。这又都说明了软件科技的发展会影响一个学科，也会将研究的方法由基本理论应用于实际。

6.2 认知理论与设计

设计与任何将观念构思、规划体现到人造物过程的心智活动有关。设计于1960年开始被研究，起始于发展出一些观念性构图或模式来解释设计过程。逐渐地，研究开始转向于研究思考过程。但经过结合许多于实验室完成的研究结果，设计中的认知终于被认证为是一个特殊的思考范围（Cross，2001），所包含的心智活动与其他专业学科，如物理、化学中的心智行为不同。例如，一些建筑师会在设计最早期发展一"**设计情节草案**"附有一些心像构成一个"解答的情境"（Chan，2001）。这种特殊的心智设计活动，在别的领域中是找不到的。因此，设计过程的研究专注于探讨设计中的认知现象。目的在于充分了解设计认知，以便改进设计质量和产品。

由解决问题的角度来研究设计认知则有两个方向可进行。一是观察设计行为，由观察中抽出类似比拟，然后把设计过程分割出一些有限的片断过程，确认每一片断过程中的变量并且找出适合每一过程的运算法，再发展出一运行这些有限片断过程的控制策略，最后化成计算机程序，完成一个类似人类思考的系统，并测试这系统。这个探讨方法是人工智能法。期待能产生一些自动设计来协助并便于设计。

另一方法是发展假说，运作实验，让受测者做一些设计课题，监看观察受测者的设计行为，或在设计工作完成后采访设计师收集原案数据，之后分析数据证据来测试假说，接着由图和口语数据中作出结论，最后做出一认知模式来验证发现。这一系列研究方法是认知科学法。目的是了解设计创造时的心理历程。这两种研究法自1970年之后做出不少结果。1990年起，使用认知科学研究设计已变成一个注目受欢迎的研究吸引点。

从设计认知的研究开始，经过实验学习后，已获得不少对设计思考的了解，也同时开启了多方面的讨论。例如，问题是如何被界定的，心像是如何被形成的，绘草图时是如何与心像结合的，注意力是如何用来控制设计过程等。这些问题仍有待未来更多的研究努力去发掘回答。在另一方面，当计算机变成设计的必需品时，做数字模的能力已和绘草图，以及做实体模一样被看成是必需的基本设计技巧之一。于是问题浮现，即有创造力的设计师如何用计算机工具提高设计能力做出一特殊设计？如果改变设计时的外在表征是否会改变心智过程？在这上下文中，外在表征呈现是指设计师用来将设计观念外显的媒体。相对的，内在表征则指的是记忆中的知识和心像的储存表征。认知实验中曾经显示出外在表征，如图和模型等，是辅助性结构，而非设计师记忆中的心像结构同形（Kosslyn，Ball & Reiser，1978）。但是，内在表征的心像运作会不会受外在表征的媒体影响？因此，如能探讨使用草图、实体模型、数字模型和虚拟模型做设计时，不同的认知状况会对思考产生影响，这将是有趣的研究课题。在此，又回到主题，即科技对认知的影响。

6.3 虚拟空间中的认知，身临其境的存在感

虚拟实景是一可用来将现实景物数字化，并在计

Artificial intelligence is a branch of computer science that studies how to endow computers with capabilities of human intelligence, and creates programs to enable computers to perform tasks or actions of an intelligent being. Artificial intelligence has been applied in many fields. It can use language rules to carry on a conversation with a human using a computer, and is even capable of speech recognition. In video games, artificial intelligence works by creating intelligence of computer controlled players or objects, so that the action of a character can react to other objects in the game. Basically, there are two purposes of artificial intelligence. One is to use the power of the computer to increase human thinking and the other is to use the computer to understand how humans think. Following these conceptual frameworks, cognitive psychology then developed and applied information processing theory, which assumes that the brain is an information processing unit. Further more, the technology of the digital computer and concepts of artificial intelligence were used as the basis to develop the field of cognitive science from cognitive psychology.

Another major branch of human cognition is human problem solving ability. Studies explored how human beings use logic and knowledge to solve particular problems. Combining the fields of cognitive psychology and artificial intelligence, a new field of **expert system** was developed. Also known as knowledge based system, an expert system is a computer program comprising a set of knowledge rules that analyze a given set of information to solve a specific problem, and provide a course of action to solve the problem. The earliest example is the software generated to deduce the molecular structure of organic components in 1965. In 1974, a practical rule-based approach to medical diagnoses created and developed a protocol for the field. Since then, the notions and methods of the expert system have been applied broadly in commercial products. This summarizes how the development of software technology influenced an academic field and changed the orientation of research from basic theory to applications.

6.2 Cognitive theory in design

Design relates to all mental activities of conceptualizing, planning, and implementing man made artifacts. Studies on design began approximately 1960 with developing conceptual diagrams or models to explain design processes. Gradually, the work changed to focus on thinking processes. It has been recognized that design cognition is a special thinking domain (Cross, 2001) that involves mental activities different from the mental behavior studied in other professional domains, such as physics or chemistry. For example, in the early stages, some architects developed a **design scenario** with certain images associated to frame the solution context (Chan, 2001). Such specific mental design activities had not been found in other domains. Thus, studies on the design process concentrated on exploring the phenomenon of design cognition. The purpose was to better understand design cognition to improve design qualities and products.

There are two ways to explore problem solving from the point of view of design cognition. One is to observe design behavior, draw analogies from observations, divide the design processes into finite processes, identify parameters and find algorithms for each finite process, develop the control structures that process the finite processes, and then finally

算机银幕上把影像以三维效果重现的新科技。进步的科技已可将现实出神入化地在银幕上逼真呈现，观众可看到周遭围绕的三维立体影像。由于可提供机会观赏计算机创出的立体环境，这环境可能是遥远看不见的世界，也可能是一个身不可及但可模拟的地点，因此许多领域已受了冲击影响。在建筑世界中，身临其境是非常有价值的设计世界表征，因为它真实地把材料的材质和空间特色一五一十地呈现出来。特别是虚拟实景的模型能将视者带到模型里去身临其境地体会内部空间。更进一步，如果虚拟实景的设施能提供模型更多的感官输入，则会产生更多的感官效果，于是一个**虚拟实境**的效果就创造出来了。

沉浸投射技术或洞穴（洞穴自动虚拟环境）是全尺度的沉浸式虚拟实景空间。它可能是10英尺或12英尺左右立方体，以半透明玻璃纤维包围的结构体。图6-1左图是三面围绕，右图是六面围绕的虚拟实景设施。影像由4台（三面围绕）、6台或48台电脑（六面围绕）产生（见附记1）投射到空间里。透过三维目镜的帮助，视者看到的是三维立体影像所环抱的世界，而显出真实感。也因为它所提供的真实性，在这世界中看到的会是熟悉的可居境地（图6-2左，北京四合院），也会是奇幻的陌生世界（图6-2右，火穴）。因此，处在这空间中所衍生出的感知经验就会是一熟悉可认的经验，也或是似曾相识的感觉。这似曾相识感可能是曾经看过或做过某事，但不一定非要在以前某一时空中曾经亲临过的经验。因此，虚拟实景是一很好的科技。军医院曾用它对越南退役军人进行心理治疗，也被用作心理外显治疗法来克服惧高症或惧飞症。

虚拟实景因为具有将人和数字世界融合成一体的优点，也被广泛用于医学手术训练、军事坦克训练、飞行训练、地理仿真、计算机游戏和博物馆中保存世袭财产等。也因此，真实的质量精度和视觉冲击两个重要因素就导引出另一**"投入存在感"**的新向量。决定在虚拟实景中**"投入存在感"**的主要变量是：影像分辨率、影像尺寸大小和观者在观赏世界中与影像之间的距离。分辨率越高，影像尺度越大，越靠近观众，则越能产生出存在感。沉浸投射的虚拟实景设施有产生投入存在感的感觉，因为观众在这环境中是被全尺度、高分辨率的影像环绕着，如身临其境。一个在六面围绕**"沉浸投射设施"**里，以原案口语法为分析方法的心理实验就经由受测者证实任何模型，不论多细附有多少材质，只要在沉浸投射中展现都会产生存在感。模型的不同也只影响投入的强弱程度而已（Chan & Weng, 2005）。这解释了存在感是虚拟实景的一项宝贵资源。

取在多面围绕全尺度的沉浸投射中，能体现存在感的优势，虚拟实景应当可映照出真实，并可依赖模拟现实来反射投映出真实中的认知，也可在模拟中测验认知。因此，在虚拟中复制现实，并在复制的现实

图6-1 三面围绕（左图）及六面围绕（右图）沉浸投射技术的虚拟实景设施（版权由虚拟实境实用中心，艾奥瓦州立大学拥有）
Figure 6-1 Immersive projection technology VR facilities of three-sided (left) and six-sided (right). Copyright by the Virtual Reality Applications Center, Iowa State University

create computer coding to complete a system similar to human thinking and test the system. This exploratory method is the artificial intelligence approach which intends to generate design automation to assist and facilitate design.

Another approach is to develop hypotheses, conduct experiments by giving design tasks to subjects and monitoring their design behavior, or interviewing designers after their design works were completed, for collecting protocol data, analyzing evidences to test the hypotheses, drawing conclusions from the drawing and verbalization, and finally developing cognitive models to verify findings. This sequential research method is a cognitive science approach to understanding mental processes of design creation. Research using these two approaches produced some results, beginning in the 1970's. By the 1990's, the application of cognitive science in the study of design thinking had become popular.

After research in design cognition began, and empirical studies were conducted, understanding in design cognition opened discussions about how problems are defined, images are shaped, sketches are connected to mental images, and how the control process is involved. These topics still require more research. As computers became more of a necessity as part of design, digital modeling became one of the essential skills, in addition to sketch drawing and physical modeling. One question emerged regarding how creative designers could use computers-based tools to enhance their ability to develop unique designs: would the mental processing in design change due to the application of different external representations? In this context, external representation means the medium used by designers for displaying design concepts, as opposed to internal representation of human knowledge and images stored in memory. Cognitive experiments have shown that external representations, such as drawings and models, are supplementary structures but not isomorphic to the designers' own image structure in memory (Kosslyn, Ball & Reiser, 1978). But, does the operation of internal representation of mental image rely on external representation of media display? It will be an interesting topic to compare the cognitive differences among the uses of sketches, physical models, digital models, and virtual models in design, which cause different impacts to design thinking. This, again, implies the phenomenon of the influence of technology to cognition.

6.3 Cognition in virtual reality, the sense of presence

Virtual reality(VR) is a new technology which can be used to digitize physical objects and three-dimensionally display them on screen. Viewers can see the surrounding images in stereoscopic form. VR has influenced many areas and provides opportunities to visualize the three-dimensional environment created in computers, which is a remote world that cannot be seen virtually, or a world that is an inaccessible but imitable place. In architecture, VR is valuable for representing the design world, as it honestly and precisely expresses the texture of materials and proportions of space. Particularly, VR models can bring viewers to the interior space of a model to experience an immersive appreciation. If the VR facility can provide the model with more sensory input yielding greater sensory effect, then a **virtual environment** is created.

Immersive Projection Technology (IPT) or CAVE (Cave Automatic Virtual Environ-

图6-2 北京四合院（左图），火穴（右图）
Figure 6-2 Beijing courtyard house (left), firecave (right)

里测验认知来了解感知是可行的方法。这方法也可测试建筑物环境来评估居住环境对居民的冲击。例如，一栋建筑是一居住空间的包含物。在包含物里给予一些环境刺激，将会引起一些认知反应。健康的环境将提供正面刺激促进有收获成效的认知反应。如此而言，办公环境中任何开放的工作空间将会比狭窄的空间更健康。因为狭窄的空间会让其中工作者有幽闭恐怖感。居民会被环境影响，是因为人心是经过视觉去体会建筑之故（Eberhard，2003）。因此，利用虚拟实景来测试在环境中的认知反应是极可能的。这一构思在两个心理实验中被试验证实过。

第一个实验（Chan，2007a）应用虚拟实境将位于美国华盛顿特区的一实际办公室数字化（见附记2及图6-3）。主要模型琢磨得像实际办公室（图6-4）。当主模完成之后，模型的材料即逐次变化，例如，办公桌加上隔间墙材料由标准的化学产品和纤维织品改成橡木、樱桃木或大理石（图6-5）。或办公桌家具、隔间墙和主要构架的材料一起改为橡木、樱桃木或大理石等等（图6-6）。十位受测者被要求在三面环绕的沉浸式虚拟实境中（图6-7），先观看并记住主模。然后受测者要对应于随后六个模型的改变，并给舒适与否从一到九的回应。九是极舒适，一是极不舒适。这实验的假设是某种材料和某种建筑形态关系密切，使用某种建材就代表这栋建筑有

某些特定的机能用途。改变建筑的材料将会改变对这建筑环境的认知。例如，木材极普遍用于不显著的建筑形态里。大理石则用于公共或公众建筑来表达庄严宏伟的感觉。如果传统的木材改为大理石材质，则心理上很难接受在半隐密空间中的新庄严变化。这现象在这实验中得到证实。亦即当受测者记住主模的空间模式之后，他们很能接受橡木和樱桃木的改变（图6-5左及中图），但对大理石材质变化感到不舒适。特别是办公桌家具、隔间墙和主要构件如果都用大理石，是会让受测者感觉舒适度最低的数字模型（图6-6右图）。

第二个实验（Chan，2007b）使用与第一个实验相同的模型，也用同样的方法测试色彩和材料对认知的差异。色彩是被认为影响人类情感和认知最有用也最有力的设计工具之一。人类会对不同颜色有不同对应，而这些对应发生在潜意识和情感层面。伯兰（1978）定义色彩有两个系统：暖色及冷色。暖色在光谱中位于红、黄之间，冷色则位于绿、紫色间。暖色会激奋人类器官；相反，冷色会舒缓、延迟人体的生理过程。就象征性而言，每一色彩都有正、反两面的意义（Nolan，2002；表6-1）。基于这些生物上的发现，可以假设空间中的红（暖）色会刺激而且振奋居民，蓝（冷）色会对居民有平静的效果。因此，这实验假设冷色比暖色在紧凑的办公空间中更能产生舒

ment) is a full scale immersive VR environment space. It could be a 10 feet or 12 feet cube of translucent fiberglass structure. Picture on the left of Figure 6-1 is a three-sided and a six-sided immersive projection technology facility on the right. Images created by four (for three-sided) and six or 48 computers (for six-sided) were projected into the space (see note 1). With the help of 3D goggles, viewers are surrounded by 3D stereoscopic images to show reality. Because of the reality it provides, it has impacted human cognition while perceiving the world which could either be a familiar livable one (see left image on Figure 6-2, Beijing courtyard house) or a strange fictional world (see right image on Figure 6-2, firecave). Perception created would either be a recognizable experience; or a "da ja vou" which is the feeling of having seen or done something before and it is not necessary to having been there in a past life. Therefore, VR is a good technology for psychological therapy used by VA hospitals to treat Vietnam veteran, or used as psychological exposure therapy to cure fear of heights or fear of flying.

Virtual reality has also been broadly applied in medical surgery training, military tank training, flight training, geology simulation, computer games, and museum display for preserving heritage due to the advantages of mixing human beings with the digital world. The quality of reality and the impact to perception are important issues, which bring up a new dimension – the **sense of presence** in the VR environment. Variables determining the sense of presence in a VR environment are the resolution and size of the image, and the distance between images and viewers in the viewing world. The higher resolution, larger image size, and closeness of the viewer create a strong sense of presence. **Immersive projection** facilities can provide a good sense of presence because viewers are surrounded by high resolution, full-scale images. A psychological experiment conducted in a six sided IPT using protocol analysis with subjects proved that any model, regardless of its resolution and quality of the material textures, as long as displayed through the IPT, would always generates a sense of presence. The differences among models only affect the degree of immersion (Chan & Weng, 2005). This emphasizes that the sense of presence is valuable in a virtual reality environment.

Owing to the advantages of the sense of presence achieved in the multiple-sided full scale IPT facility, VR could mirror the reality and reflect the cognition in reality by simulating the reality and testing cognition in the simulation. It is feasible to replicate the reality in VR, and test cognition through virtual replication to explore perception. This method could be used to evaluate the impact of a building environment on its inhabitants. For example, a building is a container for a living space. Providing environmental stimuli in a container would yield certain cognitive reactions. A healthy environment, then, should provide positive stimuli to promote productive cognitive responses. As such, in office environments, for instance, an open workspace is considered healthier than a confined space, as confined spaces tend to make most occupants feel claustrophobic. Occupants are thus affected by their environment because the human mind experiences architecture through visual perception (Eberhard, 2003). It is possible to apply VR for testing the cognitive reaction in an environment. This line of thought has been tested and substantiated in two psychological experiments.

The first experiment (Chan, 2007a) applied virtual environment to digitize a real of-

图 6-3　位于华盛顿特区的办公室空间
Figure 6-3　An office space in Washington DC

图 6-4　主要的虚拟实境数字模
Figure 6-4　The master VR digital model

图 6-5　办公桌家具加隔间墙材料是橡木（左）、樱桃木（中）或大理石（右）用作实验刺激的模型
Figure 6-5　Models with oak（left），cherry（middle） and marble（right） materials on furniture and partition walls used for experimental stimulus

图 6-6　办公桌家具、隔间墙和主要构架材料为橡木（左）、樱桃木（中）或大理石（右）的模型
Figure 6-6　Models with oak（L），cherry（M），and marble（R） on furniture，partition walls，and major structural components used for experimental stimulus

适感。随着这些假设，实验即在寻找哪一色彩会更能影响人类认知，并比较色彩和材料经认知对心理的影响。图 6-8 是在周围空间中改用红色和蓝色的两个模。图 6-9 显示色彩用于周围空间加上天花以测观者的反应。30 位受测者参与，并要求对每一看过的模型表示舒适程度以一到九的分数对应。实验结果显示蓝色的舒适度比红色稍高，但在统计上并不显著。在材料部分，橡木是最受欢迎的材料，大理石最差。总而言之，材料在统计上对于色彩在视觉认知后的舒适度而言有较高的反应。

fice located in Washington DC (Note 2 and Figure 6-2). A master model resembling the real office was constructed (see Figure 6-4). Then, materials of components were changed sequentially in the following models. For instance, furniture and partition walls were changed from standard fabric to oak, cherry, and marble (Figure 6-5); or furniture, partition wall, plus major structural components (Figure 6-6) were changed as a group to oak, cherry, and marble respectively. Ten subjects were recruited to view the master model first, in a three-sided virtual environment (Figure 6-7), and memorize it. Then they were asked to respond to the changes of materials with regard to their level of comfort ranging from one to nine. One was extremely uncomfortable and nine was extremely comfortable. The hypotheses of the experiment was that certain materials are strongly associated with certain building typologies, and the use of specific materials implies that a building has a specific purpose. Changes in the materials used in a building would change the perception of the built environment. For instance, wood is a very common building material used mostly in unremarkable contexts. Marble, however is used mostly for public and civic buildings to convey a sense of grandeur. If the traditionally wooden texture were replaced by marble, it might be difficult to psychologically accept the new environment as a semi-private office setting. This phenomenon had been proven in an experiment where, after memorizing the pattern shown in a master model, subjects were comfortable with the use of oaks and cherry (Figure 6-5 left and middle images) but not marble. The model of structural components with marble was the least comfortable model perceived by subjects (Figure 6-6, right image).

The second experiment (Chan, 2007b) applied the same model with similar procedures to test the perceptual differences between color and materials. Color is considered one of the most useful and powerful design tools to affect human emotions and perceptions. People respond to different colors in different ways, and these responses take place on a subconscious and emotional level. Birren (1978) defined two systems of color: warm and cool. Warm colors are those between red and yellow on the spectrum, whereas cool ones are between

Positive and negative meanings of color   Table 6-1

Color item	Positive meaning	Negative meaning
White	Clean, innocent, pure	Cold, empty, sterile
Red	Strong, brave, passionate	Dangerous, aggressive, domineering
Green	Natural, tranquil, relaxing	Cold, depressing, gloomy
Blue	Strong, trustworthy, authoritative	Jealous, inexperienced, greedy

色彩正、反两面的意义　　　表 6-1

颜色单位	正面意义	负面意义
白	干净、无辜、纯洁	冷淡、空虚、贫瘠
红	强力、勇敢、激情	危险、侵犯、掌控
绿	自然、宁静、松弛	冷淡、消极、阴暗
兰	强壮、信赖、权威	嫉妒、生疏、贪婪

图 6-7　六面环绕沉浸式虚拟实境中呈现的办公室

Figure 6-7　The VR office displayed in a six-sided virtual environment

这两个实验说明虚拟实境如何被用来研究对环境的认知。尤其是实际上难以做到的研究。这是因为人类是经由视觉认识这世界。而沉浸式虚拟实境是现实的翻版复制品，并且经由在虚拟世界中的投入存在感可反映出真实中的认知机能。对于未来在虚拟世界中的视觉研究，如果受测者的样本数足够，重复几次实验，能得到更显著的统计证据支持发现的结果，则会得到更有信服力的成果。因此，应用虚拟实境研究认知会是一个新的领域。这两个实验例子在另一方面也同时证实了新科技对认知科学研究方法之影响。另一例子，相同地，是另一研究领域的形成，即合并神经科学和认知心理学的"**认知神经学**"。认知神经学中带来的新科技发展和新发现会对认知科学的未来研究有极大的影响。

6.4　网络神经和网络心理学

神经科学专注于分析**神经元**的功能结构。神经元是神经系统的功能单元。在人类大脑中存在有千亿的神经元。每一单元都存有信息。因此，脑神经科学研究主掌全身系统的大脑功能；相对地，认知科学关切主掌心智活动的原则。因此，对象征性而言，神经科学是研究大脑的硬件，而认知科学研究的是大脑进行信息的软件。存在两者之间，有一居中斡旋传统认知心理学和大脑科学意图从各方面情报得到认知层次理论，包括神经元线路的运算属性、大脑受伤后被损害的行为形态，以及测量在运作某些认知事件时的大脑活动等，这新领域被称为"认知神经学"。

认知神经学的研究是探讨与认知机能有关的特殊"神经生理"活动。基本假设是说某些特定的大脑区域负责传达某些特定的认知机能（图 6-10）。"认知神经学"的研究课题，略述之，包括探讨大脑的视觉认察功能机制，运动神经的控制、学习和记忆，以及情感等。这些观念也被脑科学证实了不同的思考和脑中不同的活动形态相关。研究方法是用"**磁共振成像法**"扫描大脑来发掘血液在大脑中活跃部位的流动。磁共振成像法能逐层地扫描人类大脑，探测大脑哪些区域负责哪些认知的工作项目。例如，当朗诵句子时，脑中负责语言的部分即亮起。当在想像行走于建筑物房间时，负责空间导航和认知地点的区域即活跃起来。当想像打网球时，运动区域即被触发而投入参与。所有这些由某些神经单元操控而发生的活动事件可由扫描图中不同的颜色中观察得出。一项研究大脑进行影像活动的研究即显出十分有趣的结果（O'Craven & Kanwisher, 2000）。

在该研究中，当人们看一地方或绘一脸孔的图片时，彩色的大脑扫描图即有不同的照射热点出现。有趣的是，当人想像相同地方和脸孔时，相同的脑

green and violet. Warm colors tend to stimulate the human organism. Conversely, cool colors are relaxing and retard bodily processes. Symbolically, each color has certain positive and negative meanings (Nolan, 2002; Table 6-1). Based on these biological findings, it is assumed that red (warm) colors in a space will stimulate and excite occupants, and blue (cold) color will have a calming effect on occupants. Thus, the hypothesis in this experiment assumed that cool (blue) colors would generate more comfortable feelings in a stressful office environment than warm (red) colors did. Following these hypotheses, experiments were conducted to find out which color was more significant in affecting human psychology through perception, and to compare the impact between material and color. Figure 6-8 shows the experimental models with red and blue color applied to the surrounding walls. Figure 6-9 shows the colors also applied to the ceiling. 30 subjects participated and were required to see all models and responded with their level of comfort from scores of one to nine. Results showed that the color blue is slightly higher than the color red, but not statistically significant. In the material section, oak is most popular and marble is least popular. Overall, materials are statistically stronger than color in affecting the level of comfort after visual perception.

These two experiments explained how virtual reality could be used to study cognition in an environment which is difficult to study in reality. This is possible because human beings recognize the world through visual perception; and immersive virtual reality is a duplication of the reality which also mirrors the cognition function through the sense of presence in the virtual world. To develop a convincing argument on the value of perception in the virtual world, experimenters need enough sample sizes to repeat the experiment and obtain significant statistical results. Applying virtual reality in the study of cognition could become a new research field. These two examples of experiments also demonstrate the impact of new technology on the study of cognitive science. Another example in this regard, is an emerging research area of "**neurocognition**" which is the combination of neuroscience and cognitive psychology. New technological development and discoveries in neuroscience could impact future studies in cognitive science.

6.4 Neuroscience and neurocognition

Neuroscience concentrates on the analysis of mechanisms of **neurons**. Neurons are the functional unit of the nervous system. In our brain, there are 100 billions of neurons, each stores information. Neuroscience focuses on functions in the brain, which serves as the control center for all the body's systems. Cognitive science is concerned with principles governing mental activities. Symbolically speaking, neuroscience studies the hardware of the brain and the cognitive science studies the software of the brain regarding processing information. There is an interface between traditional cognitive psychology and the brain science attempting to gain cognitive level theories from various types of information, including the computational properties of neural circuits, patterns of behavioral damage following brain injury, and measures of brain activity during the performance of cognitive tasks. This new field is called "**neurocognition**" or "**cognitive neuroscience**".

The study of neurocognition explores the specific neurophysiological correlations of cognitive functions. The basic assumption is that specific brain regions are responsible for mediating certain aspects of cognitive function (see Figure

图 6-8 红色及蓝色用于周围墙面

Figure 6-8　Red and blue colors on surrounding walls

图 6-9 红色及蓝色用于周围墙面及天花板

Figure 6-9　Red and blue colors on partition walls and ceiling

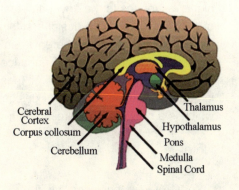 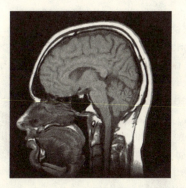

图 6-10 人类大脑图解及扫描图

Figure 6-10　Diagram and scanned image of human brain

6–10). Areas of study in neurocognition are, to name just a few, brain mechanisms underlying visual perception, motor control, leaning and memory, and emotion. Nurioscience has proven that different kinds of thoughts can be tied to different patterns of activity in the brain. This theoretical method proved this through using **Magnetic Resonance Imaging** (MRI) to scan the brain to detect blood flow to active parts of the brain. MRI is a technique that can scan the human brain layer by layer to detect the areas active while certain cognitive tasks are being performed. For instance, when reciting sentences, the parts involved in language lit up. When imagining visiting rooms of a house, the parts involved in navigating space and recognizing places activated. When imagining playing tennis, the areas that trigger motions participated. The occurrence of these activities operated by certain neuron could be seen using different colors in the scanned image. A study analyzing brain activity in relation to the processes of imagery showed an interesting result (O'Craven & Kanwisher, 2000).

In that study, colorful brain scans revealed different hot spots when people saw a place and when they saw a face. Interestingly, the same brain areas lit up when people merely imagined the same places and faces (Figure 6–11). Images on the top two rows on Figure 6–11 were seeing and imaging pictures of faces, and seeing and imaging pictures of places were shown on the bottom rows. This indicates that when people "perceive" images of faces (row 1) and places (row 3), while "thinking" about similar images of faces (row 2) and places (row 4), the same brain areas are functioning. It also reveals that certain types of information are handled by certain brain areas. It is also possible that different information is stored in different brain areas. Of course, this example explains clearly the metaphor that the human brain is the information processing unit. Applying MRI to expose thinking activities in design could be a promising method for studying fundamental brain activities while designers perform creative thinking, as a way to understand creativity, and to detect how designers use mental image and procedural knowledge for exploring design knowledge representation.

6.5 Conclusions

Design thinking involves many mental activities performed during various activities in a number of professions. The body of knowledge used by individual designers is specialized and acquired during the course of professional education and developed through professional training. Because of their specialized body of knowledge, designers have unique thinking routines that signify their individual styles (Chan, 2001); and have special views toward the environment that consequently lead the designer to react to the environment in a unique way. These concepts are based on the premise that the human mind is an information processor which automatically accepts information from the environment and applies it to specialized knowledge stored in memory, which generates unique responses to the environment. It is very important to study how designers think and how they cognitively react to the environment in order to improve design style and enhance designers' response to the environment. All these discoveries point to the significance of design cognition, which is a new academic research frontier.

Note:

1. A CAVE is enclosed by three or six walls constructed by fiberglass. Each wall has ma-

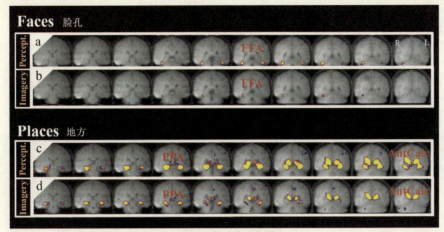

图 6-11 看和想像脸孔及地方图片时的磁共振成像法扫描图
Figure 6-11 MRI scanned images of brain activities on seeing or imaging pictures of faces and places (*Reprinted from Mental imagery of faces and places activates corresponding stimulus-specific brain regions*, by K. M. O'Craven and N. Kanwisher, p.1015, 2000, with permission from MIT Press Journals).

区域也被燃亮（图 6-11）。图 6-11 中上方两列图片是看及想像脸孔，下两列则是看及想像地方影像。这显示当人观看脸孔（列一）或地方图片（列三）跟人想像相似的脸孔（列二）或地方影像（列四）一样，由相同的脑区域运作。这解释了某种形态的情报被某个大脑区域掌控。这也诉说了不同情报确实是储存在不同的脑区域。当然，这一例子证实了人类大脑是情报进行单元的隐喻。相同地，磁共振成像法也可用来发掘设计思考活动，探讨设计师做创意思考时的根源活动，了解创造力的最佳资源，或测出设计师如何使用心像和程序知识来探讨设计知识的呈现表征。

6.5 总结

设计思考涉及许多不同专业中不同情况的心智活动。尤其是，一套由设计师个人使用的知识体系更是特殊。这套知识体系得于一系列的专业教育熏陶，以及专业训练培育而成。因为这一特殊的知识体系，设计师能有特别的思考惯例透出个人的设计风格（Chan, 2001），或有特别对环境的观点而对环境有特殊的反应。这所有的观念都根源于一基本的假设，即人心是一情报进行器，自动接受环境中的信息，利用记忆中存有的特殊专业知识，而后产生出对环境的特殊回应。当然，不言而喻，分析研究设计者如何思考以便改进设计风格，以及了解设计者如何在认知上对应环境以便加强对环境的健康互动是多么重要了。这些就是设计认知的重点。也是一个学术研究的新天地。

附记：

1. 洞穴是由三面或六面墙嵌以玻璃纤维围绕的结构体。每面墙均有机器制造出左眼和右眼两个影像投射到空间里提供三维视像。有些洞穴设施中的每面墙（包括地面）使用一台机器创出这些立体影像，也有用 8 台机器创出，每台机器负责 1/8 部分的墙面影像。

2. 实验的办公室是"能适应的工作空间实验室"，这是一联邦办公环境位于华盛顿特区联邦总务署总部大楼的 7 楼。这实验室容纳 48 位联邦雇员，总楼地板面积约为 11000 平方英尺。

chines creating left and right eye images projected into the space for a 3D view. Some facilities use one computer generating the stereographic images for one wall （plus the floor） and some use eight machines each is in charge of 1/8th portion of the wall image.

2. The office is the Adaptable Workspace Laboratory （AWL）, a federal office environment located on the seventh floor of the US General Services Administration （GSA） headquarters in Washington, DC. The AWL encompasses 11,000 square feet and houses about 50 federal agency employees.

参考书目
Bibliography

Akin, O. How do architects design? In J. C. Latombe (Ed.), *Artificial Intelligence and Pattern Recognition in Computer Aided Design*. New York: North-Holland, 1978: 65–104.

Akin, O. An exploration of the design process. *Design Methods and Theories*, 1979, 13 (3/4): 115–119.

Akin, O. *Psychology of architectural design*. London: Pion, 1986.

Akin, O., Dave, B., Pithavadian, S. Heuristic generation of layouts (HeGeL), In J. Gero (Ed.), *Artificial intelligence in engineering design*. Southampton: Computational Mechanics Publications, 1988: 413–444.

Akin, O., Chen, C. C., Dave, B., & Pithavadian, S. A schematic representation of the designers' logic, *In CAD and robotics in architecture and construction*. Paris: Hermes, 1986: 46–58.

Akin, O. & Lin, C. Design protocol data and novel design decisions. *Design Studies*, 1995, 16: 211–236.

Alexander, C. The determination of components for an Indian village, In J. C. Jones and D. G. Thornley (Eds.), *Conference on Design Methods*. Oxford: Pergamon Press, 1963: 83–144.

Alexander, C. *Notes on the synthesis of form*. Cambridge, MA: Harvard University Press, 1964.

Alexander, C., Ishikawa, S., & Silverstein, M. *A pattern language which generates multi-service centers*. Berkeley, CA: Center for Environmental Structure, 1968.

Alexander, C. The state of the art in design methods, In N. Cross, (Ed.), *Developments in design methodology*. New York: John Wiley & Sons, 1984: 309–316.

Anderson, J. R. *Cognitive psychology and its implications*. San Francisco, CA: W. H. Freeman, 1980.

Anderson, J. R. *The architecture of cognition*. Cambridge, MA: Harvard University Press, 1983.

Anderson, J. R., & Bower, G. H., *Human associative memory*. Washington, D.C.: Winston, 1973.

Archer, B. An overview of the structure of the design process, In G. T. Moore (Ed.), *Emerging methods in environmental design and planning*. Cambridge, MA: MIT Press, 1970: 285–307.

Archer, B. *Whatever became of design methodology*? Design Studies, 1979, 1 (1) :17–18.

Arnheim, R. *Toward a psychology of art.* Berkeley, CA: University of California Press, 1966.

Arnheim, R. *Art and visual perception: A psychology of the creative eye.* Berkeley, CA: University of California Press, 1974.

Atkinson, R. C. & Shiffrin, R. M. The control of short-term memory. *Scientific American*, 1971, 225: 82–90.

Atman, C., Chimka, J., Bursic, K. M., & Nachtmann, H. L. A comparison of freshman and senior engineering design processes, *Design Studies*, 1999, 20 (2): 131–152.

Birren, F. *Color and human response: Aspects of light and color bearing on the reactions of living things and the welfare of human beings.* New York: Van Nostrand Reinhold, 1978:34.

Broadbent, D. A mechanical model for human attention and immediate memory, *Psychological Review*, 1957, 64: 205–215.

Broadbent, D. *Perception and communication.* London: Pergamon Press, 1958.

Broadbent, G. Design method in architecture, In G. Broadbent, & A. Ward (Eds.), *Design methods in architecture.* New York: George Wittenborn, 1969: 15–21.

Broadbent, G. The development of design methods, *Design Methods and Theories*, 1979, 13 (1): 41–45.

Brooks, L. R. Spatial and verbal components of the act of recall. Canadian *Journal of Psychology*, 1968, 22: 349–368.

Byrne, R. W. Mental cookery: an illustration of fact retrieval from plans.*Quarterly Journal of Experimental Psychology*, 1981: 33A: 31–37.

Carramazza, A., Gorden, J., Zurif, E. B., & De Luca, D. Right hemispheric damage and verbal problem solving behavior. *Brain and Language*, 1976, 3: 41–46.

Chan, C. S. Cognition in design process " In *Proceedings of the 11th Annual Conference of the Cognitive Science Society*. Hillsdale, NJ: Lawrence Erlbaum, 1989: 291–298.

Chan, C. S. Cognitive processes in architectural design problem solving. *Design Studies*, 1990, 11 (2): 60–80.

Chan, C. S. Exploring individual style through Wright's design. *Journal of Architectural and Planning Research*, 1992, 9 (3): 207–238.

Chan, C. S. Mental image and internal representation. *Journal of Architectural and Planning Research*, 1997, 14 (1): 52–77.

Chan, C. S. An examination of the forces that generate a style. *Design Studies*, 2001, 22 (4): 319–346.

Chan, C. S. Thoughts of Herbert A. Simon – on artificial intelligence in design. In M. L. Chiu (Ed.), CAAD *talks 2: Dimensions of design computation.* Garden City Press: Taipei, 2003: 26–33.

Chan, C. S. Evaluating cognition in a work space virtually. In *Proceedings of the 12th International Conference on Computer Aided Architectural Design Reserch in Asia*, 2007a: 451–458.

Chan, C. S. Does color have weaker impact on human cognition than material?In *Proceedings of the 12th International*

Conference on Computer Aided Architectural Design Futures. Springer Science: Dordrecht, 2007b: 373–384.

Chan, C. S., & Weng, C. How real is the sense of presence in a virtual environment? In A. Bhatt (Ed.), *Proceedings of the 10th International Conference on Computer Aided Architectural Design Reserch in Asia*. New Delhi: TVB School of Habitat Studies, 2005: 188–197.

Chase, W. G., & Simon, H. A. Perception in chess. *Cognitive Psychology*, 1973, 4: 55–81.

Chase, W. G., & Simon, H. A. The mind's eye in chess. In W. G. Chase (Ed.), *Visual information processing*. London: Academic Press, 1974: 215–281.

Cherry, E. C. Some experiments on the recognition of speech, with one and with two ears. *Journal of the Acoustical Society of America*, 1953, 25: 975–979.

Chomsky, N. *Syntactic structures*. The Hague: Mouton, 1957.

Clark, H. H., & Chase, W. G. On the process of comparing sentences against pictures. *Cognitive Psychology*, 1972, 3: 472–517.

Clark, H. H., & Chase, W. G. Perceptual coding strategies in the formation and verification of descriptions. *Memory & Cognition*, 1974, 2: 101–111.

Chi, M. T. H., Glaser, R., & Rees, E. Expertise in problem solving. In R. Sternberg (Ed.), *Advances in the psychology of human intelligence*. Hillsdale, NJ: Lawrence Erlbaum Associates, 1982: 7–75.

Collins, A. M., & Loftus, E. F. Spreading activation theory of semantic processing. *Psychological Review*, 1975, 82: 407–428.

Collins. A.M., & Quillian, M. R. Retrieval time from semantic memory. *Journal of Verbal Learning and Verbal Behavior*, 1969, 8: 240–248.

Cooper, L. A., & Shepard, R. N. Chronometric studies of the rotation of mental images. In W. G. Chase (Ed.), *Visual Information Processing*. New York: Academic Press, 1973: 75–176.

Cross, N. *Developments in design methodology*. New York: John Wiley & Sons, 1984.

Cross, N. Design cognition: Results from protocol and other empirical studies of deign activity. In E. Eastman, M. McCracken & W. Newstetter (Eds.), *Design knowing and learning: Cognition in design education*. Amsterdam: Elsevier, 2001: 79–103.

Cross, N.. & Cross, A. C. Observations of teamwork and social processes in design. *Design Studies*, 1995, 16: 143–170.

Dansereau, D. F. *An information processing model of mental multiplication*. Ph.D. dissertation, Carnegie Mellon University, Pittsburgh, PA, 1969.

Darke J. The primary generator and the design process. *Design Studies*, 1979, 1 (1): 36–44.

Davies, S., & Castell, A. Contextualizing design: narratives and rationalization in empirical studies of software design. *Design Studies*, 1992, 13 (4): 379–392.

DeGroot, A. D. Perception and memory versus thought: some old ideas and recent findings. In B. Kleinmuntz (Ed.), *Problem solving.* New York: Wiley, 1966: 19–50.

Deutsch, J. A., & Deutsch, D. Attention: Some theoretical considerations.*Psychological Review*, 1963, 70: 80–90.

Dey, A. K. Understanding and using context. *Personal Ubiquitous Computing* 2001, 5 (1): 4–7.

Dorst, K. Analyzing design activity: new directions in protocol analysis. *Design Studies* Vol 16, 1995, 139–142.

Dorst, K., & Dijkhuis, J. Comparing paradigms for describing design activity, *Design Studies*, 1995, 16 (2): 261–274.

Downing, F. Conversations in imagery. *Design Studies*, 1992, 13: 291–319.

Eastman, C. On the analysis of intuitive design processes. In G. T. Moore (Ed.), *Emerging methods in environmental design and planning.* Cambridge, MA: MIT Press, 1970: 21–37.

Eberhard, J. How the brain experiences architecture, AIA *Journal of Architecture.* 2003, Spring: 1–5.

Echenique, M. Models: a discussion. In L. Martin, & L. March (Eds.), *Urban space and structures.* London: Cambridge University Press, 1972: 164–174.

Ellis, W. D. *A source book of Gestalt psychology. London* : Routledge, 1999.

Ericsson, K. A., & Simon, H. A. Verbal reports as data. *Psychological Review*, 1980, 87: 215–251.

Ericsson, K. A., & Simon, H. A. *Protocol analysis: verbal reports as data.* Cambridge, MA: MIT Press, 1984: 48–62.

Frankenberger, E., & Badke-Schaub, P. Integration of group, individual and external influences in the design process, In E. Frankenberger, P. Badke-Schaub, H. Birkhofer (Eds.), *Designers-The key to successful product development.* London, UK: Springer, 1998.

Foz, A. T. K. *Observations on designer behavior in the parti.* Master's thesis, Massachusetts Institute of Technology, 1972.

Gardner, H. *The mind's new science: A history of the cognitive revolution.* New York: Basic Books, 1985.

Gay, P. *Art and act: On causes in history–Manet, Gropius, Mondrian.* New York: Harper & Row, 1976: 123–124.

Gay, P. *Weimar culture: The outsider as insider,* New York: Harper & Row, 1980.

Gibson, E. *Principles of perceptual learning and development.* New York: Appleton-Century-Crofts, 1969.

Gibson, E., Schapiro, R., & Yonas, A. Confusion matrices for graphic patterns obtained with a latency measure. *The analysis of reading skill: A program of basic and applied research* (Final report, project No. 5-1213). Ithaca, N. Y.: Cornell University and U. S. Office of Education, 1968.

Goel, V. and Pirolli, P. The structure of design problem spaces. *Cognitive Science*, 1992, 16: 395–429.

Goldschmidt, G. The dialectics of sketching. *Creativity Research*, 1991, 4 (2): 123–143.

Goldschmidt, G. The designer as a team of one. *Design Studies*, 1995, 16: 189-209.

Guindon, R. Designing the design process: exploiting opportunistic thoughts. *Human Computer Interaction* Vol 5, 1990: 305-344.

Hartmann, G. *Gestalt psychology*. New York: Ronald Press Co, 1935.

Hasher, L., & Zacks, R. T. Automatic and effortful processes in memory. *Journal of Experimental Psychology*: General, 1979, 108: 356-388.

Hayes, J. R. Memory span for several vocabularies as a function of vocabulary size. *Quarterly Progress Report*, Acoustics Laboratory, MIT, 1952.

Healy, A. F. Proofreading errors on the word "the": New evidence on reading units. *Journal of Experimental Psychology: Human Perception and Performance*, 1980, 6: 45-57.

Heath, T. *Method in architecture*, New York: Wiley, 1984.

Hesse, M. *Models and analogies in science*. Indiana: University of Notre Dame Press, 1966.

James, W. *The principles of psychology*. New York: Holt, 1890.

Jenkins, J. G., & Dallenbach, K. M. .Oblivescence during sleep and waking. *American Journal of Psychology*, 1924, 35: 605-612.

Jones, J. C. *Design methods: Seeds of human futures*. London: Wiley, 1970.

Jones, J. C. A method of systematic design. In J. C. Jones & D. G. Thornley (Eds.), *Conference on Design Methods*. Oxford: Pergamon Press, 1963: 53-73.

Kahnemen, D. *Attention and effort*. Englewood Cliffs. NJ: Prentice-Hall, 1973.

Kant, I. *Critique of pure reason* / translated and edited by P. Guyer, A. W. Wood. Cambridge: Cambridge University Press, 1998.

Koffka, K. *Principles of Gestalt psychology*. New York: Harcourt Brace, 1935.

Koffka, K. The art of the actor as a psychological problem. *American Scholar*, 1942, 11: 315-326.

Kohler, W. Grouping in visual perception. In C. Murchison (Ed.), *Psychologies of* 1930, Worcester, Mass: Clark University Press, 1930: 143-147.

Kosslyn, S. M. *Ghosts in the mind's machine*. New York: Norton, 1983.

Kosslyn, S. M. Mental imagery. In D. N. Osherson, S. M. Kosslyn, & J. M. Hollerbach (Eds.), *Invitation to cognitive science*. Cambridge, MA: MIT Press, 1990: 73-97.

Kosslyn, S. M., Ball, T. M., & Reiser, B. J. Visual images preserve metric spatial information: Evidence from studies of image scanning. *Journal of Experimental Psychology*: Human Perception and Performance, 1978, 4: 47-60.

Kosslyn, S. M., & Pomerantz, J. R. Imagery, propositions, and the form of internal representations. *Cognitive Psychology*, 1977, 9: 52–76.

LaBerge, D., & Samuels, S. J. Toward a theory of automatic information processing in reading. *Cognitive Psychology*, 1974, 6: 293–323.

Lea, G. Chronometric analysis of the method of loci. *Journal of Experimental Psychology: Human Perception and Performance*, 1975, 104: 95–104.

Levy, B. A. Role of articulation in auditory and visual short-term memory. *Journal of Verbal Learning and Verbal Behavior*, 1971, 10: 123–132.

Lindsay, P. H., & Norman, D. A. *Human information processing: An introduction to psychology.* New York: Academic Press, 1977.

Lloyd, P., Lawson, B., & Scott, P. Can concurrent verbalisation reveal design cognition? *Design Studies*, 1996, 16 (2): 237–259.

McNeill, T., Gero, J., et. al. Understanding conceptual electronic design using protocol analysis. *Research in Engineering Design*, 1998, 10 (3): 129–140.

Miller, G. A. The magical number seven, plus or minus two: Some limits on our capacity for processing information. *Psychological Review*, 1956, 63: 81–97.

Miller, G., Galanter, E., & Pribram, K. *Plans and the structure of behavior.* New York: Holt, Rinehart and Winston, 1960.

Miller, G. A., & Isard, S. Some perceptual consequences of linguistic rules. *Journal of Verbal Learning and Verbal Behavior*, 1963, 2: 217–228.

Mook, D. *Classic experiments in psychology*, Westport, CT: Greenwood Press, 2004: 177–182.

Moray, N. Attention in dichotic listening: Affective cues and the influence of instructions. *Quarterly Journal of Experimental Psychology*, 1959, 11: 56–60.

Neisser, U. *Cognitive psychology.* Englewood Cliffs, NJ: Prentice Hall, 1967.

Newell, A. Heuristic programming: Ill-structured problems. In J. Aronofsky (Ed.), *Progress in operations research* (pp. 360–414). New York: Wiley, 1969: 360–414.

Newell, A., Shaw, J. C., & Simon, H. A. Elements of a theory of human problem solving. *Psychological Review*, *1958, 85: 151–166.*

Newell, A., & Simon, H. A. *Human problem solving.* Englewood Cliffs, NJ: Prentice-Hall, 1972.

Nickerson, R. S. A note on long-term recognition memory for pictorial material. *Psychonomic Science*, 1968, 11: 58.

Nolan, K., *Color it effective: How color influence user*, 2003. http: //office. microsoft. com/en-us/assistance/HA010429371033. aspx.

Norman, D. A., & Rumelhart, D. E. *Exploration in cognition.* San Francisco, CA: Freeman, 1975.

Norman, D. *Memory and Attention: An introduction to human information processing.* New York: John Wiley & Sons, 1976.

O'Craven, K.M., & Kanwisher, N.K. Mental imagery of faces and places activates corresponding stimulus-specific brain regions. *Journal of Cognitive Neuroscience*, 2000, 12 (6): 1013-23.

Paivio, A. *Imagery and verbal processes.* New York: Holt, Rinhart and Winston, 1971.

Paivio, A. *Mental representations: A dual coding approach.* New York: Oxford University Press, 1986.

Palmer, S. E. The effects of contextual scenes on the identification of objects. *Memory and Cognition*, 1975, 3: 519-526.

Palmer, S. E. Hierarchical structure in perceptual representation. *Cognitive Psychology*, 1977, 9: 441-474.

Perkins, D. *Knowledge as design.* Hillsdale, NJ: Lawrence Erlbaum Associates, 1986.

Peterson, L. R., & Peterson, M. J. Short-term retention of individual verbal items. *Journal of Experimental Psychology*, 1959, 58: 193-198.

Phillips, W. A. On the distinction between sensory storage and short-term visual memory. *Perception & Psychophysics*, 1974, 16: 283-290.

Piaget, J. *The psychology of intelligence*, London: Routledge and Kegan Paul, 1967.

Piaget, J. *The equilibration of cognitive structures: The central problem of intellectual development.* Chicago: University of Chicago Press, 1979.

Porter, T. *How architects visualize.* New York: Van Nostrand Reinhold, 1979.

Posner, M. I. Abstraction and the process of recognition, In G. H. Brower (Ed.), *The psychology of learning and motivation.* New York: Academic Press, Vol. 3, 1969.

Posner, M. I. *Cognition: an introduction*, Glenview, IL: Scott, Foresman, 1973.

Rapoport, A. *House form and culture.* Englewood Cliffs, NJ: Prentice-Hall, 1969.

Reed, S. K. Structural descriptions and the limitations of visual images. Memory and Cognition, 1974, 2: 329-336.

Reitman, J. Without surreptitious rehearsal, information in short-term memory decays. *Journal of Verbal Learning and Verbal Behavior*, 1974, 13: 365-377.

Reitman, J. S. Skill perception in Go: Deducting memory structures from inter-response times. *Cognitive Psychology*, 1976, 8: 336-357.

Reitman, J. S., & Rueler, H. H. Organization revealed by recall orders and confirmed by pauses. *Cognitive Psychology*, 1980, 12: 554-581.

Reitman, W. R. Heuristic decision procedures, open constraints, and the structure of ill-defined problems. In M. W. Shelley & G. L. Bryan (Eds.), *Human judgments and optimality.* New York: Wiley, 1964: 282-315.

Rock, I. *The logic of perception.* Cambridge, MA: MIT Press, 1983.

Rowe, E. J. & Rogers, T. B. Effects of concurrent auditory shadowing on free recall and recognition of pictures and words. Journal of *Experimental Psychology: Human Learning and Memory*, 1975, 104: 415–422.

Rowe, P. *Design thinking*. Cambridge, MA: MIT Press, 1987.

Rubin, E. *Synsoplevede figuere*. Copenhagen: Gyldendalsde Boghandel. Translated into German (1921) *as Visuell wahrgenommene figuren* (same publisher), 1915.

Rumelhart, D. E. Schemata: the building blocks of cognition. In R. J. Sprio, B. C. Bruce, & W. F. Brewer (Eds.), *Theoretical issues in reading comprehension*. Hillsdale, NJ: Lawrence Erlbaum Associates, 1980: 33–58.

Rumelhart, D. E., & Ortony, A. The representation of knowledge in memory. In R. C. Anderson, R. J. Sprio, & W. E. Montague (Eds.), *Schooling and the acquisition of knowledge*. Hillsdale, NJ: Lawrence Erlbaum Associates, 1977: 99–135.

Rundus, D. Analysis of rehearsal processes in free recall. *Journal of Experimental Psychology*, 1971, 89: 63–77.

Scheidig, W. *Weimar crafts of the Bauhaus, 1919–1924: An early experiment in industrial design*. New York: Reinhold, 1967.

Schon, D. A. *The reflective practitioner*. London, UK: Temple-Smith, 1983.

Schon, D. A. Designing: rules, types and worlds. *Design Studies*, 1988, 9 (3): 181–190.

Shepard, R. N. Recognition memory for words, sentences, and pictures. *Journal of Verbal Learning and Verbal Behavior*, 1967, 6: 156–163.

Shepard, R. N. The mental image. *American Psychologist*, 1978, 33: 125–137.

Shepard, R. N., & Metzler, J. Mental rotation of three-dimensional objects. *Science*, 1971, 171: 701–703.

Simon, H. A. *Models of man*. New York: Wiley, 1957.

Simon, H. A. *Sciences of the artificial*. Cambridge, MA: MIT Press, 1969.

Simon, H. A. Style in design. In J. Archea, & C. Eastman (Eds.), *Proceedings of the 2nd Annual Environmental Design Research Association Conference*. Stroudsbury, PA: Dowden, Hutchinson & Ross, 1970: 1–10.

Simon, H. A. The structured of ill-structured problem. *Artificial Intelligence*, 1973, 4: 181–201.

Simon, H. A. How big is a chunk? *Science*, 1974, 183: 482–488.

Simon, H. A. The functional equivalence of problem solving skills. *Cognitive Psychology*, 1975, 7: 268–288.

Simon, H. A., & Barenfeld, M. Information processing analysis of perceptual processes in problem solving. *Psychological Review*, 1969, 76: 473–483.

Simon, H. A., & Lea, G. Problem solving and rule induction: a unified view. In L. W. Gregg (Ed.), *Knowledge and cognition*. Potomac, MD: Lawrence Erlbaum Associates, 1974: 105–127.

Singer, E. Raising consciousness. *Technology Review*, 2007, 110 (1): 50–54.

Sobel, C. P. *The Cognitive sciences: An interdisciplinary approach*. Mountain View, CA: Mayfield Publishing Company, 2001.

Sperling, G. A. The information available in brief visual presentation. *Psychological Monographs*, 1960, 74 (11, Whole No. 498): 1–29.

Sperling, G. A. A model for visual memory tasks. *Human Factors*, 1963, 5: 19–31.

Standing, L. Learning 10, 000 pictures, *Quarterly Journal of Experimental Psychology*, 1973, 25: 207–222.

Sternberg, S. Two operations in character recognition: Some evidence from reaction time measurements. *Perception & Psychophysics*, 1967, 2: 45–53.

Stiny, G. & March, L. Design machines. *Environment and Planning B: Planning and Design*, 1981, 8: 245–255.

Susanuma, S. Kanji versus Kana processing alexia with transient agraphia. *Cortex X*, 1974, 10: 89–97.

Suwa, M., Purcell, T., & Gero, J. Macroscopic analysis of design processes based on a scheme for coding designers' cognitive actions. *Design Studies*, 1998, 19: 455–483.

Teuber, M. L. New aspects of Paul Klee's Bauhaus style. In M. L. Teuber (Ed.), *Paul Klee: Paintings and Watercolors from the Bauhaus Years*, 1921–1931. Des Moines Art Center, 1973: 6–17.

Teuber, M. L. Sources of ambiguity in the prints of Maurits C. Escher. *Scientific American*, 1974, 231: 90–104.

Thomas, J. C., & Carroll, J. M. The psychological study of design. *Design Studies*, 1979, 1: 5–11.

Townsend, J. T. Theoretical analysis of an alphabetic confusion matrix. *Perception & Psychophysics*, 1971, 9: 40–50.

Treisman, A. M. Contextual cues in encoding listening. Quarterly Journal of *Experimental Psychology*, 1960, 12: 242–248.

Turing, A. A. Computing machinery and intelligence. *Mind*, LIX (236), 1950: 433–460.

Valkenburg, R., & Dorst, K. The reflective practice of design teams. *Design Studies*, 1998, 19: 249–271.

Visser, W. Use of episodic knowledge and information in design problem solving. *Design Studies*, 1995, 16: 171–187.

Wade, J. *Architecture, problems, and purposes*. New York: Wiley, 1977.

Wallas, O. *The art of thought*. New York: Harcourt Brace Jovanovich, 1926.

Warren, R. M. Perceptual restorations of missing speech sounds. *Science*, 1970, 167: 392–393.

Warren, R. M. & Warren, R. P. Auditory illusions and confusions. *Scientific American*, 1970, 223: 30–36.

Watson, J. B. *Behaviorism*. New York: Norton, 1924.

Wertheimer, M. Untersuchungen zur Lehre von der Gestalt II. in *Psycologische Forschung*, 1923, 4: 301–350.

Translation, entitled Laws of Organization in Perceptural Forms, published in Ellis, W. (1999). *A Source Book of Gestalt Psychology* (pp. 71–88). London: Routledge, 1999.

Winograd, T. W. Frame representations and the declarative-procedural controversy. In D. G. Bobrow and A. Collins (Eds.), *Representation and understanding: studies in cognitive science*. New York: Academic, 1975: 185–200.

Wundt, W. *Principles of physiological psychology.* translated by Edward Bradford Titchener. London: S. Sonnenschein and Co, 1910.

Zeisel, J. *Inquiry by design: Tools for environment-behavior research*, Monterey, CA: Brooks/Cole Publishing Co, 1981.

中英对照检索
Chinese-English Translation Index

Algorithms	算法	p. 103, 106
Arts and Crafts Movement	英国美术工艺运动	p. 23, 24
Association	联结（或联想）	p. 8, 9, 55, 56
Attention	注意力	p. 39, 40
Bauhaus, School of Art and Craft	包豪斯设计学院（包豪斯）	p. 23, 24
Behaviorism	行为主义心理学	p. 7, 8
Breadth-first search	广度优先搜寻	p. 87, 90, 92
Case-based reasoning	案例式推理	p. 125, 126, 139, 140
Chunk	组集（或组块）	p. 55, 56
Cocktail party phenomenon	鸡尾酒会现象	p. 39, 40
Coding	资料代码	p. 63, 64
Cognitive mechanism	认知机制	p. 14, 15
Cognitive psychology	认知心理学	p. 10, 11
Cognitive science	认知科学	p. 12, 13
Conceptual-driven process	观念导向过程	p. 39, 40, 45, 48
Concurrent verbalization	同步叙述法	p. 105, 106
Context-aware ness	情境感知	p. 121, 122
Critical problem situation	关键性的问题情况	p. 122, 123
Cybernetics	模控学（或神经机械学）	p. 31, 32, 35, 36
Data-driven	数据导向	p. 39, 40, 45, 48

Declarative knowledge	陈述性知识	p. 127, 128
Depth-first search	深度优先搜寻	p. 85, 90, 91, 92
Design cognition	设计认知	p. 21, 22, 97, 98
Design constraint	设计约束	P.122, 123, 139, 140
Design scenario	设计情节草案	p. 143, 144
Developmental psychology	发展心理学	p. 10, 11
Divide-and-conquer	分裂而征服	p. 103, 106
Domain specific knowledge	领域知识	p. 65, 68
Effector	效应器	p. 65, 66
Encoding	编码	p. 69, 70
Episode	部曲	p. 110, 111
Experimental psychology	实验心理学	p. 10, 11
Experimental Stimuli	实验刺激	p. 51, 52
Expert system	专家系统	P.99, 100, 143, 144
Extraneous variable	外在变量	p. 73, 76, 95, 96
Field theory	场学说	p. 21, 22
Functionalism	机能主义心理学派	p. 7, 8
Heuristics	启发诱导式	p. 103, 104
Holistic	全体宏观	p. 43, 46, 59, 60
Hypothetical reasoning	假设性推理	p. 103, 106
Gestalt psychology	完形心理学（或格式塔心理学）	p. 8, 9, 21, 22
Global constraint	全域约束	p. 125, 126
Go	围棋	p. 73, 76
Ill-defined (ill-structured) problem	非明确界定问题（弱构问题）	p. 99, 100
Immersive Projection Technology	沉浸投射技术	p. 145, 146
Information	信息	p. 12, 13
Information theory	信息论	p. 12, 13
Information processing theory	人类信息处理学（或信息论学）	p. 12, 13, 37, 38
Input attention	输入注意力	p. 41, 44
Introspective verbalization	内省口语法	p. 7, 8, 103, 106
Latency	等待时间	p. 135, 136
Law of Area and Symmetry	区域与对称律	p. 25, 36
Law of Closure	结束律	p. 25, 36
Law of Common Fate (Uniform Destiny)	共同命运（单一终点）律	p. 25, 26

Law of Figure/Ground	图形与地景的分离律	p. 26, 27
Law of Good Continuation	好的连续律	p. 25, 26
Law of Pragnanz	简单完好律	p. 23, 24
Law of Proximity	接近律	p. 23, 24
Law of Similarity	相似律	p. 23, 26
Local constraint	局部约束	p. 125, 126
Long-term memory (LTM)	长期记忆	p. 37, 38
Means-ends analysis	手段－目的分析法	p. 103, 104
Magnetic Resonance Imaging	磁共振成像	p. 151, 154
Memory span	记忆广度	p. 53, 54
Mental image	心智影像（心像）	p. 47, 50
Mental picture	心智图像	p. 61, 62
Mental set	心套（或心向作用）	p. 103, 106, 139, 140
Mental structure	心智结构	p. 14, 15
Mental processing	心智过程	p. 14, 15
Mnemonics	记忆增进术	p. 57, 58
Network theory	网络理论	p. 55, 58
Neuron	神经元	p. 151, 152
Neurocognition	认知神经学	p. 151, 152
Operational research	运筹学（或作业研究）	p. 29, 30
Parti	原基	p. 35, 36
Partial-report procedure method	部份回报程序法	p. 41, 44
Pattern language	模式语言	p. 30, 31
Pattern recognition	形态辨认	p. 43, 46
Perception	知觉	p. 37, 38, 39, 40, 69, 72
Perceptual organization laws	视觉组织规律	p. 23, 24
Perceptual-test	视觉测试	p. 121, 122
Pre-solution model	先决模型	p. 125, 126
Problem Behavior Graph	问题行为图解	p. 111, 114
Problem solving theory	解题模式理论	p. 29, 30, 97, 98
Problem space	问题空间	p. 99, 102
Problem structure	问题的结构	p. 101, 104
Procedural knowledge	程序性知识	p. 127, 128
Protocol analysis	原案口语分析	p. 105, 108

Protocol statement	原案陈述	p. 109，110
Reaction time	反应时间	P. 47, 50, 65, 66, 69, 70, 109, 110
Recognition	辨认	P. 37，38
Rehearsal	复诵	p. 55，56
Representation	表征（呈现表示、或重现表示）	p. 37, 38, 61, 62, 101, 104
Retrieval	提取	p. 69，70
Retrospective verbalization	追忆口语法	p. 105，106
Rules of thumb	大拇指法则（或经验法则）	p. 103，104
Schema	基模（或知识模式）	p. 10, 11, 63, 64, 127, 128
Semantic network theory	语意网络理论	p. 55, 58, 63, 65, 66, 68
Sense of presence	投入存在感	p. 145，148
Sensory buffer	感觉缓冲存储器	p. 37，38
Sensory register	感觉收录器	p. 37，38，51，52
Short-term memory (STM)	短期记忆	p. 37，38，53，54
Social psychology	社会心理学	p. 10，11
Solution context	解答的情境	p. 125，126
Stimulus-response	刺激和反应	p. 8，9，12，13
Structuralist	结构主义心理学派	p. 7，8
Tachistoscope	塔驰斯投镜	p. 53，54，59，60
Think out loud	口语出声法	p. 73，76，107，110
Transferring	转移	p. 69. 70
Trial-and-Error	尝试-错误法	p. 103，104
Turing Test	图灵测试	p. 141，142
Ubiquitous Computing	遍布运算（无所不在运算）	p. 121，122
Verbal coding	文字码	p. 75，78
Virtual reality	虚拟实景	p. 143，146
Virtual environment	虚拟实境	p. 145，146
Visual coding	视觉码	p. 75，78
Well-defined (well-structured) problem	明确界定问题（良构问题）	p. 99，100
Whole-report procedure	全体回报法	p. 41，44